MW01352752

FARMING
Without Losing Your Hat

A Practical Guide to the Brutal Realities of the Best Job in the World

BY PAUL DORRANCE

Foreword by Will Harris of White Oak Pastures

Farming Without Losing Your Hat

Acres U.S.A.
P.O. Box 1690
Greeley, Colorado 80631 U.S.A.
(970) 392-4464 • (800) 355-5313 U.S./Canada
info@acresusa.com • www.acresusa.com

Printed in the United States of America

Edited by Ben Trollinger.

Foreword by Will Harris.

All images used with permission.

Publisher's Cataloging-in-Publication
Dorrance, Paul, 1979-
Farming without losing your hat / Paul Dorrance. Greeley, CO, ACRES U.S.A., 2022, 214 pp., 26 cm.
Includes index, tables, photographs and illustrations

ISBN 978-1-60173-170-8 (trade)

1. Farm management. 2. Record keeping. 3. Marketing. 4. Production standards. 5. Farm work.

I. Dorrance, Paul, 1979-
II. Farming without losing your hat.
III. S560.0 S571.5

To Caleb, Carissa, and Caitlin;
whose mere presence in my life compels me to be a better man,
father, and farmer.

Acknowledgements

This book is written in an effort to educate, encourage, and equip those who dare to dream of engaging in a higher, cleaner, and healthier form of agriculture. Taking that terrifying, yet extremely gratifying, step into the unknown is the best thing I've ever done. My humble efforts on behalf of those who would join me have been shepherded by a small but dedicated group of supporters.

To my family, thank you for cheering this effort and egging me on when it seemed unattainable.

I am grateful as well for my Acres U.S.A. family, who guided me in both my farming and writing efforts. To Ryan Slabaugh, Sarah Day Levesque, Ben Trollinger, Cameron Ehrlich, Hannah Fields, Keri Hayes, Chad Kuskie, Paul Meyer, Jessica Smith, and Rachel Wobeter: thank you all from the bottom of my heart. Without your support, this book does not exist.

I owe a huge debt of gratitude to Angela Blatt, who served as initial editor for this book. Far more importantly, Angela is my sounding board, cheerleader, counselor, and confidante. It is a rare thing indeed to find someone who can lift you up, yet provide a much-needed dose of reality when required. My advice: hold those people particularly close and never let them go.

Finally, thank you to my friend and hero Will Harris, for showing the world how raising animals for food should be done.

Contents

A Foreword / A Bit of Advice

By Will Harris, White Oak Pastures
Bluffton, Georgia

I'm not one to watch my language. We are a profane and irreverent people, but I'm going to try to be good here. I need to make this one really important thing clear without using expletives. I don't believe in "How-To" books about farming and ranching. They are a bunch of bull.

Complicated operations, like factories, lend themselves to "How-To" manuals. They work in that application. But you cannot operate a complex operation, like White Oak Pastures, with a manual. It is entirely too situational. Don't buy a book that you intend to use as a road map to take you to where you want your farm to be. Buy a book that will help you to think about your farm differently.

When I began rethinking White Oak Pastures, I had three minimum wage employees and I was selling a little more than a million dollars of live cattle each year. Today, we have 180 employees and we're selling twenty-something million dollars of products a year. What'd we change to get to that point? Pretty much everything. We exited the industrial, commoditized, centralized agriculture model. In doing that, we internalized all of the functions that Big Ag had previously provided for us. We internalized the costs, too.

When I was part of the industrial model, I ran the farm the way my father did. I focused exclusively on the monocultural production of a commodity beef operation. I kept my business records in a size 11-D Redwing boot box. Every January, I pushed the box across the table to my accountant and waited to see how much taxes I would have to pay. We always made money in that system. We always had a little money in the bank, and we had no debt.

But after a few years I found that I was enjoying this production system less and less. I did not like the unintended consequences of that model. The impact on my land, and my animals, and my community were displeasing to me.

We were not working with nature to produce the best products that we could. We were producing product that would just meet the minimum USDA standards. The excesses of our production system increasingly bothered me. I had come to recognize the unintended consequences of the tools that reductionist science had given us to take costs out of production. These undesirable—and previously unnoticed—consequences fell on the welfare of the animals, the degradation of the land and in the impoverishment of rural America.

This was in the mid-1990s. I had also begun to read about sophisticated consumers who wanted to eat food raised differently. So I decided to transition my farm, and that started with a new business approach, a different operational structure, and better land and animal management. The first thing I did was to stop using performance enhancing drugs or hormone implants and sub-therapeutic antibiotics on my animals. We stopped continuous grazing on monocultural pastures (if you don't remember anything else, remember this—nature abhors a monoculture). I quit using chemical fertilizer and tillage and pesticides on my land. I made the decision to quit feeding grain to ruminants. And I thought that was all I ever had to do. And I thought I was the premier, born-again farmer.

Eventually, I realized that real business practices needed to evolve and progress in the same way my growing practices were changing. We could not dump our product into the commodity market. This market would not allow us to extract the added value that we had put into the product. So we borrowed $7.5 million to do our own processing. Suddenly we had debt, and I learned the hard way what operating at a loss felt like. At this point, we weren't making money anymore.

One day I woke up and we had 100 employees and everyone worked for Will. I had put no management structure in place. I determined the financial well-being of my company every Friday night, I would pour a big glass of wine and pull up my bank account balance. I would usually go, "I believe that's all right." But a few times I said, "Oh sh**."

But I changed all of that. First, I brought in the best people that

I could hire. I had to figure out housing for these employees in creative ways. We live in a very rural area, and that continues to be our biggest challenge. I later brought in my family, and they found new ways to diversify, to construct this organism we are managing today. So we built our own eCommerce store, we started a restaurant, we built a RV Park, we invested in a dozen cabins for agritourism, we erected an open community pavilion, we opened a General Store, we made biodiesel fuel, we sold and made soap, we grew multiple species of ruminants and poultry, we tanned hides, we made pet chews—do you know how much folks will pay for pet chews?—we built out our grocery distribution, and we even produced black soldier flies for chicken feed. All in all, we now operate a bunch of businesses that make a little money, or lose a little money, but they create the entire organism that is White Oak Pastures. We've failed in some areas too — make no mistake about that.

I'm going to shut up now and let you get on with it. This is just a little bit about our farm, and what we do, and that mindset I thought I might share with you. There are a lot of other ways to do it. You might find a co-op is your best format, but that's not for me. I was an only child and have always struggled to "work and play well with others." You will have to figure out what works for your specific context.

I could pontificate more about all this, but it would be at the risk of selling you a "how-to" path and misleading you that there is such a thing. To be clear, there is no guaranteed path to success in this field. Your path will depend on your personality, your ecosystem, your market, the money you have to start with, the dream you have in your head, and the goals you set from the first day you commit to becoming a farmer or rancher. Your path will be much different than mine.

So read this book. Use it to build your own, original idea for what you can do. Stay independent, and find the time, money and management to do what you think that you need to do — you will need all three — to create your dream farm. And don't buy into any silver-bullet system that tells you how to farm. That's the entire adventure you're about to embark on — you, and only you, can figure that out.

I wish you good luck.

Will Harris is a fourth-generation cattleman, who tends the same land that his great-grandfather settled in Bluffton, Georgia. After Will took over managing the farm he made the bold decision to transition the pesticide-, herbicide-, hormone- and antibiotic-intensive farming system to a regenerative farming system like his grandfather had 130 years before. Will and White Oak Pastures have been recognized all over the world as leaders in humane animal husbandry and environmental sustainability.

Introduction

If you are reading these words, then I already owe you a debt of gratitude. You are serious enough about your future farm journey to take a concrete step toward making your dreams a reality. By allowing this book—and, by extension, me—to become a part of your journey, you are helping me fulfill my own dreams of who I want to be in the world of regenerative agriculture. Thank you for that honor!

Whenever I speak to a group, I'm quick to point out that I am not a farmer by trade. I didn't grow up in an existing farm business, a fact that has both hobbled me with costly mistakes and freed me from the bounds of conventional thinking. I grew up "country" in upstate New York, the firstborn son of two hippies who found their way back to the land. Growing up on 55 acres in New York's Southern Tier dairy country, I had all the benefits of a homestead Guernsey milk cow, homegrown beef, a small laying flock, several horses and a goat named Blossom—all financially supported by two parents with full-time jobs. It was a blessed upbringing, to be sure.

After graduating high school, completing a four-year Bachelor of Science degree, and commissioning into the United States Air Force as a 2nd Lieutenant, I began my professional career in earnest: flying military aircraft in service to this great country and its citizens. For the next 12 years, agriculture was the furthest thing from my mind. Not only that but, despite my childhood ties to the land, I still fell into the trap of taking food for granted. To be honest, for most of my adolescent life (in which I include college and several years after, where I mostly still acted like a kid) I didn't give food a second thought. It wasn't until the impending arrival of my first child in 2009 that I underwent "The Awakening."

As any new parent will attest, bringing a child into this world is a profound business. Also terrifying. No longer are you permitted to selfishly think only for yourself, tell yourself the same lies regarding what you put into your body three times a day, or cling to your old vices… instead you almost subconsciously seek out ways to improve yourself, this time for the benefit of your child who cannot yet help themselves. In my case, food took center stage for the first time in my life. No longer was I satisfied buying from a nameless, faceless, just-in-time food system plagued with synthetic chemicals, genetically modified organisms, and diminishing nutritive value. Instead, I began to seek out local, organic, and grass-fed foods. In short, for the first time in my life, I began to value food.

To my son Caleb: for this truth, and for the innumerable blessings that have followed since your birth, I am grateful.

Fast forward a few more years, after which active-duty military and I were clearly done with each other, I decided that it was time to put my money where my mouth was. With twelve years of military service under my belt, it was time for a grand new adventure. And so, in August of 2013, I purchased 111 acres in southcentral Ohio and Pastured Providence Farmstead was born. For the next seven years I successfully and profitably produced grass-fed beef and lamb, non-genetically modified pastured pork and poultry, and free-range eggs.

As I got started in farming, there was a distinct lack of trustworthy, reliable, replicable information to guide me. The first few years of my farming career was through the School of Hard Knocks, and not just in the operational arena. It became very clear that most of the guidance that did exist centered on management: how to grow, graze, feed, plant, harvest, treat, and breed. Even less concentrated on the less sexy but arguably more important aspects of how to brand, market, sell, monetize, track, expense, and profit from my hard work. I resolved to do something about that problem at my earliest convenience.

In the irony of ironies, I would never have had the time to write this book if I was still farming. The elephant in the room is that Pastured Providence Farmstead no longer exists, not because it wasn't successful or profitable, but because of my failed marriage. I want to be clear here: I didn't lose the farm in the divorce, but my divorce put me in a position that I couldn't continue to farm all by myself. And so, I made the heart-wrenching decision to sell all my animals, make high-quality hay on my pastures, and pursue another dream to educate, encourage, and equip both new and seasoned farmers for success in pasture-based animal agriculture.

Through my consulting work with the newly branded Pastured

Providence *(pasturedprovidence.com)*, educational speaking opportunities, and now this book, my goal is not to tell my story as a prescription for success but instead to help you write your own success story. Please understand... I'm no expert. I'll just point out all the mistakes I made, in hopes that you'll be able to avoid some of the same and save your energy for the mistakes you'll make on your own. My stories and advice are based in boots-on-the-ground experience, not PhD-level dissertations. There is a time and a place for both, but my role here is one of counselor, advisor, and cheerleader. I desperately want you to succeed in making your own farm dream a reality.

To that end, I want to ask you for a favor: please don't buy this book and put it on a shelf. It isn't meant to be a pretty knickknack; it is meant to be used. Scribbled in. Highlighted. Carried about. Drawn on. There is extra space around all the margins so that you can make notes or jot down your thoughts as they come to you. There are open pages at the end of each chapter that have some thought-provoking questions to answer, but also space for you to capture your reactions, initial impressions, and early aspirations. Settle into that space after reading a chapter and be deliberate about answering the questions I ask, but also encapsulate your own questions, answers, and dreams. I truly believe that it is within those pages that the value of this book will reveal itself to you. As Will Harris mentions in his foreword, the purpose of this book is not for you to follow the path I took to success in farming; instead it is for you to write your own success story.

There is extra space around all the margins so that you can make notes or jot down your thoughts as they come to you.

Now more than ever, the stage is set for ecologically minded farms and farmers to be successful. Consumer demand for clean, healthy, humanely produced food has never been higher. Opportunities for new and beginning farmers have never been so accessible. Agriculture is in desperate need of new blood and innovative approaches, but in order to capitalize on all of these opportunities there is one thing we need most of all: we need farmers to be profitable through running a thriving farm business. Without the business concepts in place, your farm will never live up to its full potential. You must set your farm on a firm business foundation, and above all else that is what this book aims to help you do: farm without losing your hat!

— *Paul Dorrance*

CHAPTER 1

What Brings You Here?

I'm going to go out on a limb and guess that you aren't considering getting into farming for the paycheck. Call me crazy, but I'd even put money down on that bet. When I sat in my rocking chair on Columbus Air Force Base back in 2012, half-listening to the training planes droning overhead and wishing I were somewhere else, I wasn't daydreaming about rolling around in a pile of money like Scrooge McDuck. Most of the time what was on my mind, and what I'm guessing is on yours, was some assortment of gorgeous sunrises, frolicking animals, rainbows, the smell of fresh-cut hay, hard yet honorable work, eating peas and beans straight off the vine, walking your land in the evening, gorgeous sunsets, and falling asleep to the soft, subtle chirping of crickets—only to wake up the next day and experience it all again.

The Lies We Tell Ourselves

While I'm obviously being a little facetious, I'd be remiss if I didn't say that I have experienced all of those things in my farming journey and they are genuinely amazing. I've never had them all happen in the same day, but I guess that is one of the things that keeps me going; maybe today is the day! Regardless, my point is this: in those rocking chair moments, while we settle into our current house/job/situation/relationship believing that something better awaits—when we allow our minds to drift toward all of the sunrises, rainbows, and sunsets that could be ours someday—we are lying to ourselves. Point blank. Straight to our own faces. Lies.

Does that sound a little harsh? Perhaps, but I wish that someone would have said that to me eight years ago. Not because it would have changed my future or my decision to farm. No, I'm far too belligerent and ornery to have allowed it to change my made-up mind. But I wish I would have heard that message back then for one reason: to better prepare myself for what was waiting for me in between the sunrises and sunsets, to better prepare myself for the realities of farming. When I talk to people who want to get into agriculture, this is how I describe my early farming experience: "The highs are as amazing as I had hoped they would be, but the lows are so much lower than I thought."

Farming is tough, and I was 100 percent guilty of dreaming through rose-colored glasses. I overlooked, generalized, or outright ignored the level of stress, mental trauma, and difficulty this lifestyle presents. Think about it: if it was easy then everyone would do it! Who wouldn't want to spend their day on the land, earning their living by the sweat, determination, and individual effort that are the hallmarks of farming? Who wouldn't want to see a concrete, real, non-digitized result from their work at the end of a long day, before falling into bed with the feeling of satisfied exhaustion knowing that something actually got done? This way of life absolutely offers those things, but you're going to earn it!

The Highs Are High

My daughter was only two years old when we moved to the farm, but she quickly displayed a natural desire and empathy for the animals that surrounded her. She often asked to accompany me for chores, which I was generally more than happy to do. She slowed me down and made me a lot less productive, but at the time I was smart enough to recognize those moments for what they were: an opportunity to shape and develop her young mind. She was my little farm buddy. At one point, walking hand-in-hand with her across the yard on a bright, sunny summer day, I noticed our shadows on the ground. One tall shadow reaching down, one short shadow reaching up, joined together in common purpose. I took a picture of those shadows with my phone, and to this day it is one of my favorite photos and fondest memories. Life was good and my heart was full... the high was indeed high.

Later in the year, it was time to process a batch of meat chickens. Because Ohio allows a certain amount of birds to be processed on-farm without inspection, I had jumped into that option for my broilers and turkeys. I hung kill cones on a post, which allow a bird to be calmly held upside down and secure while their jugulars are cut and they bleed out, peacefully slipping away into meat heaven. It is an important task and one that I place a high importance on getting "right," but it is also one of the most distasteful parts of my job. In the midst of processing day my daughter joined me outside, providing a welcome distraction from the grim task at hand. As I walked past her carrying a chicken from the holding pen to the kill cone, she stopped me to ask, "Daddy, can I please pet the chicken?" Caught by surprise, I defaulted to my normal answer when my children ask to participate in anything farm related: "Of course."

Kneeling down, I held the doomed bird while my sweet little girl stroked its feathers, cooing and talking sweet to it, telling it how pretty it was. All I could think of was the next step in this interrupted trip, and how to shield my baby from what was about to happen. In the end, I just told her that the chicken had to go and asked her to go back into the house, then stood up and walked around the corner to the kill cone. Concentrating on ensuring a humane kill, I flipped the chicken upside down and carefully placed it into the cone, gently grabbing its head as it popped out of the bottom opening. Two firm swipes of my knife and the bird's lifeblood flowed out onto the ground; it was done.

Then and only then did I recognize the feeling of being watched, and my heart dropped. My daughter hadn't gone into the house after all, but instead had followed me around the corner and witnessed the entire process. A thousand thoughts ran through my head. How was I going to explain to her that I had taken the life of the bird that she had just been petting and loving on? Would she be scarred for life by my deceit? Did I just create a two-year-old vegan? How was she going to react?

Slowly turning, I took in the scene and prepared myself for the worst. She stood there mouth wide open, eyes as big as saucers, and I could almost feel her disbelief, judgment, and confusion. Then all of my concerns fell away as she looked at me, then the dead bird, then back at me, exclaiming "COOOOOOOL!!!!" Never again did I underestimate my children's capacity to care deeply for animals yet understand that they existed for the purposes of sustaining us. This experience would shape her outlook on food and agriculture in a positive way for the rest of her life, and I am so grateful that it worked out that way... the high was indeed high.

But the Lows Are Really Low

If you choose to raise livestock, it is a given that animals are going to die. When I first started my farm, I thought I had a pretty good handle on that reality. The part that I forgot to tell myself was that more often than not, the animals weren't going to just keel over and pass away in one last blaze of glory. Instead, they were much more likely to get sick, have physical deformations, or suffer a traumatic accident that would result in their not being viable. I think everyone would agree that the humane thing to do in those scenarios, when the hurt or pain or injury is not fixable, is to end their suffering by ending their life. Yeah...no one prepared me for that part.

This is coming from a guy whose only real experience killing an animal was hunting deer from relatively long range with a rifle. Through a scope, immersed in the mental calculations of windage, elevation and bullet drop, the skill of shooting and making a clean, humane kill tended to crowd out the realization that I was ending an animal's life. When the shot was off and the animal dropped, the emotions took over, but at that point the deer was already making that all-important mental transition from "animal" to "meat."

I grew up on a homestead in upstate New York and can remember a situation that happened shortly after Dad had put some rodent poison out in the barn. I was walking over to the barn and came across a small mouse that had gotten into the poison and was clearly dying, wandering around in circles in the barnyard in broad daylight. I went to get a shovel, intent on doing the right thing and putting it out of its misery, but when I got there...I couldn't! It was too personal, too up close, and I am ashamed to say that I scooped that little mouse up in the shovel and threw it out into the hay field to die a slow painful death. But at least I didn't have to watch it happen or, more importantly to this story, participate in the process; how selfish of me.

Fast forward to my own farm, and I found that I struggled with the same issue. It didn't happen often, but when an animal got sick or injured I wrestled internally with the same conflict. Do I let the process take its natural course while the animal clearly suffered, or do I step in and end the suffering?

One beautiful April morning a pair of lambs were born, and for whatever reason one of them came out with back legs that simply didn't work. The poor thing would prop itself up on its two front legs and try to nurse, but its back legs had no strength in them and stretched straight out behind it. The lamb dragged itself around the pen for 30 minutes or so, calling out to its mama in hunger and frustration, while I steeled myself for what I needed to do next. That answer was clear to me, but the method in which to carry it out was not. I cataloged my options:

- Shoot it—No, too messy; what if I miss or cause more pain?
- Club it—No, too risky; what if I don't do it right and cause more pain?
- Stab it—No, too messy, plus painful for the lamb
- Drown it—Hmm... quick, clean, painless, quiet

Anyone reading this who has been around livestock for a while is already shaking their head in dismay, but that's my entire point. I was put into this highly emotional, time sensitive, unexpected, and traumatic situation without any warning; and if you choose to raise livestock then someday so will you. I chose what I thought best in the moment and decided to drown the lamb.

A five-gallon bucket of water later and there I was: submerging this hours-old lamb underwater to end its pitiful life. And it was quiet. And clean. And I truly believe, mostly painless. But it was NOT quick. A minute passed, and the lamb still kicked in my hands. Another minute passed and its eyes were still open, wide with panic at not being able to breath, struggling to get to the surface as I continued to hold it underwater. At that point I knew that I had messed up, that this was not the right choice, but I couldn't go back now. I had to stay there, feeling this lamb fight for its life, watching it gasp for air that I wouldn't give it, finally and mercifully relaxing into death after more than three minutes. The low was so incredibly deep and low.

I completely lost my mind, weeping and crying in the middle of the barn, while in the background the ewe gently called for her baby that I had just tortured. I'll never forget that moment. I'll never forgive myself. This tragedy was not in the narrative that I had told myself back in the comfort of my rocking chair. Nowhere was "make poor decisions and contribute to the suffering of the animals under your care" part of my farm dream. Nonetheless it had happened to me and in some form or another, fellow farm dreamer, it will someday happen to you. It is less a question of "what will I do in that moment" than "how do I pick myself up" and "where do I go from here." Incidentally, from that moment on, I decided to end life on my farm with a .22 bullet to the head, and both my animals and I have been well served by that method.

Passion Is a Requirement

Some might be wondering why on earth I decided that this chapter should be at the front of the book instead of the back. Why not save this touchy, dark, difficult topic matter for last? Am I trying to scare you away from starting your own farm? Not exactly. If these stories make you reconsider whether farming is really for you, then my job is done. This way of life is really, really difficult, both physically and mentally, and there is no need to waste your time learning about branding your business or rotational grazing if the lifestyle isn't for you. There is no shame in coming to that conclusion—just lots of heartache and angst avoided. Lend this book to another farm dreamer and start thinking about what is next for you outside of animal agriculture.

If, however, you are still reading, then congratulations—you might have what it takes to thrive on your future farm. For you, I started with this topic to drive home the importance of finding a way to maximize the highs and overcome the lows. You can do this, you can learn from your mistakes, and you can shape your future farm into an enterprise that mirrors your dreams. But in order for that to happen, above all else, you must have one thing: passion.

More important than any knowledge, experience, or marketing savvy, if you aren't passionate about agriculture and your method of farming, then you simply can't survive for long.

Too many times, farmers get caught up in an internal argument on what makes their farm "sustainable." That word means something different to everyone, but is often used in conventional agriculture to simply mean "economically sustainable." In theory, I don't disagree with that approach. In fact, much of this book is dedicated to helping you position your future farm for financial success. But as is usual in monoculture thinking, the scope of that definition is far too narrow. If you are reading this book simply so that you can run a profitable farming enterprise, then please put it down and walk away. Passion, and the commitment that is associated with it, is the one thing that made my farm and me sustainable through the inevitable lows.

As I dreamed about my future farm and got started building it from scratch, that passion manifested itself in a desire to produce good food for others, a belief in the mandate for humane animal care, and the challenge to be part of an agricultural solution to ecological issues. Not to make a dollar, not to be economically viable (although those things are critically important), but to treat my animals well, steward my land, and serve my family and community by providing them healthy, clean food—that is what drives me. That is what gets me through the hard times. That is what stops me from quitting. There are thousands of other jobs that pay so much more and aren't nearly as challenging as agriculture. Only passion kept me from throwing my hands up, walking away, and taking the easy way out.

So I encourage you to dream big; it's the only way to get started on the journey toward your farm. However, as much as possible, recognize and prepare yourself for the fact that your farm dream won't exactly match your farm reality. The good times are amazing, but the bad times are truly dark. Develop and test your passion, and if it doesn't really lie in farming, be smart enough to walk away. Ask yourself the question, "what brings me here," and be truthful with yourself in your answer. If your answer and passion lie in farming, then my goal is to help you to succeed, and to build the farm of your dreams into the farm of your reality!

Write Your Own Story

When you close your eyes and dream about your future farm, what do you see? Where are you, what are you raising/growing, and who is working alongside you? What passions are driving you forward toward the decision to farm?

CHAPTER 2

Testing the Waters—Options for Self-Education

Hey, you're still here! That's amazing, because Chapter 1 is a tough read. It made my 12-year-old son cry when I related the story to him the other day, and I'd be lying if I didn't tear up a little myself as I replayed the situation in my own mind. Putting that on paper did serve as a not-so-subtle reminder of the importance and implications of my decisions as a livestock manager, and the concrete repercussions for when I mess it up. That on its own makes it worth calling up the ugly memory on occasion. I'm still not sure of the wisdom of putting that chapter first in this book, but since you're reading it now, it must be too late to change my mind! I hope you got the point: that if you don't have a strong passion for this kind of farming, then it is not likely that you will succeed in the long run. If you still think you want to dive into starting your own farm after that, then I say "let's get started!"

Read Voraciously

When the idea that I might leave my secure, high-paying job as an Air Force pilot to start a farm really started to take hold, I found myself voracious for information. I'm sure it is the same for you. All of a sudden my Amazon Christmas list was overrun with book listings like Joel Salatin's *You Can Farm*, Lisa M. Hamilton's *Deeply Rooted*, Forrest Pritchard's *Gaining Ground*, and Gene Logsdon's *All Flesh Is Grass*. I would get up early and stay up late consuming information. Little did I know that I was already forming my belief structure, farming methods, and general approach to food and agriculture—long before I would take action on anything concrete.

Some of the things I read excited me in their simplicity, others challenged me with their complexity, and still others concerned me with their rigidity. I definitely didn't agree with all of the resident experts of the time, nor do I expect you to agree 100 percent with what you'll read here. The first step in success for your operation is to soak up as much information and book knowledge as you can, from as many sources as possible. Read, read, read...and if you can't read very well, get the audiobook or listen to a podcast. My challenge at the beginning is for you to absorb as much information and as many ideas as possible, mostly because when you actually start farming your free time disappears in a hurry!

Treat it like school, because it is. Take notes, make a reference sheet for ideas that peak your interest, highlight and tab your books. Be methodical, intentional, and committed. And don't concentrate just in your perceived niche. For example, we livestock ranchers could learn a lot from the concepts in Louise Riotte's companion-planting book, *Carrots Love Tomatoes*. While books are wonderful in their targeted and specific offerings, don't forget about the varied, yet shorter and (sometimes) more timely, information offered by magazines. As I dove deeper into this world from the comfort of my couch, I subscribed to *Acres U.S.A.*, *The Stockman Grass Farmer*, and *Farming Magazine* and found them all equally informative and thought provoking in their own ways.

You might notice that I am leaving out many of the electronic and digital options that are available to us for self-education, and that is mostly on purpose. Admitting that it could be because of the way my brain works, I would argue that mass-generated emails, list-serves, and e-magazines do not hold the same level of value as physical books and magazines. The level of study that I took in my journey and that I am advocating for you requires the ability to underline, highlight, scribble, and doodle and to flip an actual page. If you are facing a long airline flight and have an electronic version of a book on your Kindle, then great...go nuts. But the information you are soaking up now needs to be referable, recallable, and researchable once you start your farm, and for that me that means hard-copy books.

Before I leave this topic, I'd be remiss if I didn't encourage you to take a break from your classwork occasionally. Relax, enjoy yourself, and recognize that for all the scientific and operation-oriented literature that you should be immersing yourself in, you also need to take a little time for pleasure. My version of this truth found me enjoying Michael Pollan's *The Omnivore's Dilemma*, John Collis' *The Worm Forgives the Plough*, Jenna Woginrich's *Cold Antler Farm*, and E.B. White's *One Man's Meat*. One of the things that separates sustainable and regenerative agriculture from mainstream conventional approaches is our embracing of the reality that what we are doing is much more art than it is science. Yes, we need to pursue scientific methods, replicable studies, and documented results. But heaven forgive us if we don't heed the words of the great Wendell Berry when he challenges convention: "We have neglected the truth that a good farmer is a craftsman of the highest order, a kind of artist."

Agricultural Conferences

Along with self-directed book learning, there are lots of opportunities to hone your knowledge by connecting directly with existing farmers. I have found that agricultural conferences, while notably more expensive than buying a book, are an amazing way to add to my volume of information. Here you can connect not only with experienced speakers, scientists, activists, farmers, and food advocates, but perhaps more importantly you get the opportunity to meet other farmers from all over your state, region, and country—folks just like you, or a little further along than you, and some who are where you were a year or two ago. Among this wealth of knowledge and experience, you have the chance to explore specific questions, find targeted answers to your most burning issues, and establish long-lasting relationships that will serve you well into the future.

Sound exhausting? Well, especially to us introverts, it is. But it is so very helpful, and I cannot recommend attending one strongly enough. Take them in small doses, though, perhaps picking one conference to be your "staple" and maybe another to rotate around and explore different niches. Don't be afraid to mix it up; often the content can be similar in some arenas and you can mitigate that by switching up your conference schedule occasionally. The best annual conferences that I found for myself were Acres U.S.A.'s Eco-Ag Conference, which rotates around the country every year, and the Ohio Ecological Food and Farm Association (OEFFA) conference, which centers on the Midwest region and states surrounding Ohio. I recently enjoyed the Carolina Meat Conference, which is held in various cities in North Carolina and is a very specific and targeted gathering around niche and pastured livestock production, processing, and distribution. To my shame, I have yet to make it to a Grass-Fed Exchange or a MOSES conference, but they are absolutely on my list.

Whichever conference you choose to attend, make the most of your financial investment. Plan to get up early and stay out late. Attend a session every window of opportunity, and spend your free time approaching speakers to ask specific questions or introducing yourself to a random stranger that you think has an operation similar to what you envision for yourself. Very few times in our distancing society is it not only acceptable, but desirable, to find an open seat at a table full of random strangers, introduce yourself, and ask "what do you farm?" Again, treat it like school, and have your pen and notepad ready. Take lots of business cards if you have them.

But what if you aren't comfortable introducing yourself to random people and engaging in conversation? Honestly, I'll tell you to get over it for a weekend. Fake it if you have to. Assume an alter-ego. I don't care how it happens, but my earnest belief is that the true value of attending a conference isn't in the academic sessions or the keynote addresses; it is in the conversation that happens during the coffee breaks, lunch windows, and after-hours hotel bars. Trust me—I'm an introvert at heart and conferences wear me out. When I get back to the farm, I want nothing more than to engage in some sort of manual labor, surround myself with the natural environment, and not see another person for at least a week! Even so, the information gathered and relationships formed during an annual conference are incredibly important to your success in farming.

Online Courses

One of the negative aspects of conferences is the money spent on travel, food, drinks, and lodging. Lately, there is a trend developing within the agricultural community to develop and offer deep-level online coursework. While there is usually a cost associated with the classes, it pales in comparison to the amount of money spent to get yourself to a conference. You just let the conference come to you! Obviously, you lose out on the social and interpersonal interactions offered by conferences, as well as the depth and breadth of topics available in one place, so don't misunderstand me to say that you shouldn't go in person. However, keep your eye out for online agricultural classes that might suit your needs. For example, I am currently working with Acres U.S.A. to offer a course titled "The Business of Pastured Livestock Production," and there will almost certainly be others as well!

The beauty of this new type of offerings is that you can "attend" the sessions at your convenience and on your timeline. Homeschooling your kids? No problem—the coursework is available in the mornings before they wake up or in the evenings after they go to bed! Work third shift down at the plant? Easy fix; just log on and catch up on your studies after you've come home and taken a nap! It is difficult to overstate the convenience of this method of self-education, and I strongly recommend that you find some time amidst your busy schedule to enroll in and reap the benefit of these new online courseware offerings. They can vary in their cost, time requirements, and subject matter, so you will need to do a little extra work to ensure that you are seeking out courses that target your desired learning objectives. Perhaps soon, as more and more of these courses pop up, some entrepreneur will see the need for a clearinghouse of sorts that will catalog topics, costs, and time restraints to identify all the available options in one place. Maybe that entrepreneur is you!

As you can tell, it is my strong opinion that, despite recognizing the value offered by online courses, there really is something special that happens in the act of going and doing. This applies to conferences, yes, but even more importantly to what I hope you will consider as your next step in self-education before jumping into your own farming business: an honest-to-goodness, in-person "farm school."

Farm School

While I was still over two years away from making the jump into farming, my wife Heather and I attended a weekend farm school in Elberton, Georgia. At the time, their operation was almost exactly like what I envisioned mine to be. They rotationally grazed cattle and sheep, supplemented pastured hogs with the nutritious byproducts from their small grass dairy, raised laying hens and turkeys on pasture, and had rabbits and geese in addition to their cheese-making business. Needless to say, they had a very cool farm! The couple that ran the farm direct-sold their products to customers in the Atlanta area, including multiple restaurants who paid top dollar for their amazing products.

If you know anything about my farming journey, you'll see all of the obvious and exciting parallels between that operation and mine. The husband was clearly a driven entrepreneur, capitalizing on and fulfilling the rising demand for pastured, wholesome, and healthy food. He was also an astute businessman, with a lot of advice and information to offer on running a successful business, marketing, customer interactions, and telling your story...all things that would turn out to be absolutely critical in the success of my operation. Plus they offered a two-day farm school at their place. Perfect!

Despite the fact that Heather was 7 months pregnant with our second child, we registered and attended. And we had a blast. It was a perfect weather weekend, and met other couples from Tennessee, Kansas, and Texas who were equally committed to starting their own farms. We learned about everything from "search-engine optimization" to creating your business' "elevator speech." We were served excellent, wholesome meals made from on-farm and local products. We were challenged to put our dreams onto paper, so that we could ultimately make those dreams a reality. Most importantly, we were situated on an actual farm and had the opportunity to see different enterprises in person and in action.

There is something distinctly different between reading about rotational grazing concepts versus actually going out and building fence, moving animals, shifting water troughs and observing livestock behaviors. Perusing advice on porcine behaviors just doesn't give you the same level of healthy fear as stepping into a pen of hungry hogs with a bucket of whey or grain. This "boots on the ground" education is critical to translating theory to practice, for subjecting an uninformed, general excitement for livestock husbandry to the realities of animal behaviors, and for testing your commitment to your future farm. Attending a farm school should be clarifying, motivating, and challenging, all at the same time. The real value represented here is that you get to experience a very small yet representative taste of true farming, but ultimately get to leave when it's over. How important it must be to gain that experience and then be able to step back into your established, unscrambled, stable world while you consider whether or not to upend everything to start a farm.

This concept takes on a multitude of forms, including but not limited to a formal school offered by a pastured livestock producer, an Armed To Farm event targeting veteran beginning farmers, or an in-depth grazing workshop offered by groups like Practical Farmers of Iowa or Stockman Grass Farmer. Regardless of what you find for yourself, make sure that there is a hands-on portion of the program. Sitting in a classroom learning about grazing theory is good, getting out on the land to attend a pasture walk or see a farm in operation is even better. Again, there is something special about getting out of presentation mode and putting literal boots on the ground.

As Heather and I approached our farm school experience in Georgia, it was really she that needed convincing. True to my personality, I had made my decision and was all-in from the beginning, a trait that has both helped and harmed me many times over in my life. Heather was not hindered with such a trait and was hesitant to sign off on leaving the comfort of a steady job and spending our life savings on this harebrained farming scheme. To her credit, she attended and participated despite being ridiculously pregnant, and ultimately had a blast. At the end of the weekend, as we walked through the sheep paddock amidst the bouncing lambs and gazed out over the varied enterprises and healthy, satisfied animals represented on that property, Heather looked at me and said those six magic words that I will never forget:

"I think we can do this."

And just like that, we were off to the races and I never looked back.

Internships

If attending a farm school isn't enough to put you firmly on one side of the fence or the other, there is another educational opportunity for those who really need to dive into the realities of farming before forging their own way. I never personally pursued it, not because it isn't an extremely valuable experience—it just required more time than I personally wanted to spend...I was ready to get started! Farm internships represent a formal on-farm learning environment that immerses the student in day-to-day farm life. You essentially become a temporary hired hand, helping with and experiencing both the monotony and excitement that chores, seasonal variability, and animal (mis)behavior offers. Typically, internships run for at least a season, and some more in-depth opportunities last up to a year and resemble an apprenticeship model of learning.

There are several online listings for available internships, but my favorite one is through the National Center for Appropriate Technology's (NCAT) ATTRA program: *attra.ncat.org/internships*. You can narrow your search by keyword, state or with a visual map.

When you are considering an internship, recognize that it not only represents a significant increase in educational opportunity, but also a relational intensification. Internships aren't passive; they are participatory, and you will be all up in your host farmer's professional and personal space. In part, that is by design...how else are you to learn the ebbs and flows, pacing, decision-making processes, and timing requirements? With that said, it opens up a next-level can of worms for both the farmer and the intern. You'll both need to be conscious of the very real threat to overstep boundaries, step on toes, and generally be more harm than help at the beginning of the learning curve. Hopefully many of the danger zones will be discussed before any work is done but, just like any relationship, communication will be key. I suppose that's really the point—in an internship environment, you have crossed the line from employee to personal relationship, and the weight of that reality shouldn't be overlooked or understated. As the intern, your intentions are presumably pure, but your execution will be awkward at the beginning, and that is both understandable and acceptable. Your farmer will need to grant you grace to allow for those lessons to be learned.

Make sure that you give that same grace, however, recognizing that your farmer is investing in you, counting on you, and opening themselves to the highly likely scenario where your mistakes, questions, and rookie moves put them in a real pinch. Hopefully nothing gets broken and no one gets hurt, but both of those are real possibilities as you learn by doing within the construct of an internship program.

Don't Ever Stop Learning

Although this chapter is about options for self-education prior to starting your own farming operation, it is worth mentioning that your agricultural education should never stop. It will certainly get harder to prioritize as you involve yourself in the day-to-day operation of your own farm, but my challenge to you is to keep reading, keep attending conferences, and keep taking online courses. Too often we humans get settled in the rut of routine, and it benefits us to be shaken up, challenged, and reset occasionally. Keep an open mind, continuously consider alternative options and ideas, and be intentional about periodically looking over your operation with a fresh set of eyes.

Finally, be sure to pay it forward. After a few years, maybe you will begin speaking at conferences, sharing your experiences and expertise with the next generation of new and beginning farmers. Maybe you can host a farm school of your own, contributing that level of hands-on education to those who need it. Inevitably, as your operation grows, you will need an extra set of hands, representing a prime opportunity to build an internship program that fills your workforce needs while providing an opportunity for someone else to build their knowledge before starting their own farm. Writing, speaking, teaching—whatever it looks like, make sure to keep learning and educating yourself, in addition to investing in the others who will undoubtedly be coming behind you. In this way, regenerative agriculture will remain strong, vibrant, and innovative as we continue to build "craftsmen of the highest order."

Write Your Own Story

Go to your bookshelf...what farming topics are represented there already, and are there any obvious gaps in the subject matter? What other educational options would you benefit from at this stage of your journey? Can you identify any ruts that you've already developed and need to get out of?

CHAPTER 3

The Business Plan, Part I—Setting the Foundation

After taking the time to educate yourself, it is time to turn your attention to educating those around you. I'm not talking about your built-in support network—those who would love you and cheer for you even if you were contemplating leaving your steady day job to join a nudist colony in Alaska. I'm talking more about those who will need a little time and a lot of convincing to come around to this cockamamie scheme you are planning—those who will think that starting a pasture-based livestock operation and marketing your products directly to consumers is crazier than your nudist colony idea. Think bank lenders and Farm Service Agency bureaucrats. They won't be able to see your dream, much less help you fund it, unless you are able to put some concrete ideas and numbers on paper. Hence, the need for a business plan.

The other person that benefits from the admittedly painful yet critically important process of writing a business plan is you. This is ultimately where all of your amazing ideas and rainbow dreams of gorgeous sunrises and frolicking animals gets turned into reality, or at least a roadmap pointed toward that reality. As you work through the process of capturing your values, developing your vision, crafting your mission, and setting goals, something equally as exciting begins to happen. Your dreams begin to take shape—likely not exactly as you envision right now, but a better version of them. Better because these dreams are realistic. They are fully researched and supported by reality. And they represent the success that is waiting for you within your soon-to-be-established farm enterprises. Now that is exciting stuff!

Before we get there, let's be sure to firmly and purposely put the horse in front of the cart and start this process with a discussion about our values.

Values Must Come Before Vision

By definition, farms are—or, more accurately, should be—treated as businesses. In one form or another, farmers produce a product, bring or send it to a market, and get a paycheck in return for their toils. So doesn't it make sense that economics, the Almighty Dollar, and the financial bottom line should call the shots? I've heard it said that "if it doesn't make money, it doesn't make sense," and to a certain point I agree. However, I do challenge the preeminence of that way of thinking, and believe that it should—no, it must—be subservient to a concept that is more important, more powerful, and more compelling: your values.

As we embark on this journey to define, discover, and detail your future farm, it is critical to start by asking a very important question: "what type of life do you want to lead?" This isn't the only question you will ask yourself as you move through this book, but it is undoubtedly the most important. The reason is simple... agriculture, more than any other occupation that I am aware of, has the ability to test our mettle and challenge our preconceptions. As we've already discovered in Chapter 1, this lifestyle is tough. And the old Aaron Tippin song had it right when it said "you've got to stand for something, or you'll fall for anything."

There are plenty of amazing moments and seasons cradled within this lifestyle—please don't forget that part! However, in my experience, it is when the literal manure hits the fan that I need to be reminded of my values. That makes sense, right? We don't typically "need" our values or our bedrock principles when things are going great. Generally speaking, our values become critical in the heartache, in the trauma, and in the emergency, and it is important to recognize that farming has way more than its fair share of all of those.

I used to "joke" with friends that I would wake up in the morning on a gorgeous and sunny day, stretch long and hard, and wonder to myself: "What emergency is the farm going to throw at me today? Will it be a sick animal? Maybe a predator running rampant? Ooh, I know...I'm planning on baling hay, so it will definitely be some sort of broken machinery." Then, after coffee and morning chores, I would head back toward the house, only just seeing the geyser of water venting itself out of the ground..."ahh, ruptured water line, I didn't see that one coming!" But don't miss my point: I never doubted it was coming. Sometimes I woke up to the emergency, sometimes I went to sleep exhausted after dealing with it, but it was always lurking. Who else but farmers would sign up for this kind of perpetual stress and difficulty?

One of the unique and enjoyable things about farming compared to other occupations is that we generally get to call the shots and make the decisions. Of course, Mother Nature and the Good Lord always have the final say, but more than other jobs, we often get to chart our own path. It is likely one of the attributes that draws you to this lifestyle! The trap here is that there is usually no one watching when we are tempted to cut corners, to take the easy way out, or to cheat the system. Without—or sometimes in spite of—our values, this represents a real struggle.

Here's a concrete example: One spring early on in my journey, I was wrestling with the reality that my cattle were dealing with mites—little critters that cause skin irritation, hair loss, and stress. The cows were patchy, ugly, and ornery; although I suppose I would have been too if I were them! The "normal" and "easy" way to combat mites is to use a chemical pour-on that kills pretty much anything:

fleas, ticks, lice, intestinal worms, and, yes, mites. It is also incredibly harmful to the environment, gets excreted in their manure, and kills earthworms, dung beetles, and a multitude of beneficial critters in your pasture. Even knowing that, I succumbed to the lie that it was too much trouble to dust the cattle with diatomaceous earth (an effective organic topical treatment) and too expensive to run out and buy a scratching post to allow the animals to remove the mites themselves. Instead, I went against my values, treated the herd with Ivermectin, and *in one moment of weakness* I single handedly wiped out every dung beetle on my property for the next 3 years.

After that, my values formally included that I would never use a product that isn't specifically authorized by the USDA Organic Standards. I'm was not and have never been certified organic, but this was an easy way for me to draw a line in the sand, communicate value to my customers, and elevate my values above the temptation of the easy way out.

The only way we farmers can stand in the face of the isolation, risk and tragedy that is part of our day-to-day life is to cling to our values. And they have to be identified and codified before they are needed. In the middle of the difficulty is the wrong time to determine what we stand for; we humans are too weak and fallible for that! When the cattle were in the chute that spring day was the wrong time to decide if chemical treatments were right for me. With that in mind, I am going to challenge you to think through, define and refine your values now, before you need them. In fact, these values will be referenced throughout this chapter to guide and direct your business planning. We're going to dream, cast a vision and talk big picture, but then use that big picture to create concrete goals and a plan. But from the very beginning we have to realize that those goals and plans must be built on a foundation of values. Only then can we farm to our full potential, communicate accurately to our customer base, and mitigate the troubling times that farming will certainly bring our way, so we can get back to concentrating on the exciting and enjoyable times!

Capturing Your Values

Capturing your values can be done through a multitude of options, the most critical thing is that it makes sense to you. Feel free to write an essay, craft short bullet points, or draw pictures. The important part is to be intentional about identifying and capturing your values. Often, just asking "What are your values?" feels overwhelming somehow, so many times this task can be broken up into more manageable subsections. For example: Personal, Economic, Environmental, and Community. Again, go with what makes sense to you and your situation.

When I started my farm, one of my personal values read: "I want to raise animals in a more natural environment than is conventionally done." Another said: "I value family time and want to create a business where I can work together with my son." One of my environmental values was: "I desire to honor God through strong stewardship of the land that He gives me." There is no rush here; I wrote those values back in the spring of 2011, two-and-a-half years before I finally started the farm and when I only had one of my three kids!

This is a good time to discuss the likely reality that you aren't in this thing alone. If you are married, your spouse has to not only be on board; their input has to be considered. Trying to start a business that your spouse isn't also committed to is a fool's game. Ask me how I know. Perhaps you and a friend are considering partnering on an enterprise, or you have an investor that is helping you get started. Whoever you would count on your team needs to be a part of this process and work through this together with you. After you've got a decent list of values built, take some time to sit down with your team and identify the shared values that emerge. You don't have to be 100 percent matched; in fact, you almost certainly won't be. But the values your team shares across the board represent your core strength, vision, and power. Knowing where values differ is just as critical and will help you shape future goals and tasks that your entire team can support. Save the information you collect and notes you make, as all of it will inform and flow into eventually creating your farm business plan.

Developing Your Vision

Now we get to the fun part—or at least it was for me! This is the part of the process where you get to dream big, think grand, and fill your mind and heart with all the good feelings that brought you here. Instead of thinking about where you are (although much of this will be necessarily based there), now you are free to imagine your ideal farm: what it looks like, who it serves, and how it operates. It really is heady stuff, the kind of stuff that will get you through tough times and remind you of why you made this crazy decision to start a farm in the first place.

Dreaming about what you want your farm operation to look like in the future (whether that be one year from now or ten), and capturing that on paper, will help establish a destination to which you can create a pathway through appropriate planning. But for now, it's time to turn off your left brain and turn loose the right side. We won't leave organization or logical steps too far behind—I'm wound far too tight for that—but, as much as possible, let your imagination run wild for the moment. Don't worry as much about reality, specifics, goals, or strategies. Rather, your goal is to capture your farm dream and vision. To help do that, here are some questions you can ask yourself:

- What are you producing and how?
- Where and to whom are you selling your products?
- What other (intrinsic) values do you get from or share on your farm?
- Who else is involved with your farm?

Don't forget to go back and look through your values as a reminder of what you wrote. In fact, feel free to continue referencing them throughout this process—that's exactly what they are there for! But be sure to open yourself to new ideas and concepts as well. This exercise is also about brainstorming.

When I completed this process for the first time, my vision statement looked like this:

> My farm dream is to have a sustainable farm where I can produce enough food for my family and others. I want healthy, nutritious food that I can be proud of and excited about. My passion is healthy food without unnecessary chemicals, a happy place for my family, and enjoying the animals. I hope to gain life satisfaction and gratification, knowing I am helping the environment and helping other people to become healthier. My kids are involved, growing up on a place where the land is beautiful, expansive, and apart from the world.

That actually brings tears to my eyes, as I remember capturing that dream on paper and thinking to myself "this might actually be possible." I hope the same happens for you, as you create your own vision statement and enjoy the chance to dream big!

Mission Statement

A mission statement is intended to capture the essence of your business: why it exists, who it serves, and how it functions. It encompasses your guiding principles, guides both your overall strategy and day-to-day decisions, and communicates the value that you offer to customers, investors, and potential partners.

This is a perfect opportunity for me to call out the differences with the process we are using here to develop a mission statement versus what one might typically do without a structured approach. If someone asked you, "What are you thinking about doing on your farm?" I'm betting that you tend to start with the concrete. Something like, "Well, I'd like to rotationally graze beef and sell it direct to customers." Then you are at risk of starting with that very specific goal and try to build up from there, likely pushing aside other considerations such as the farm as a whole, your family, and your existing market conditions. To avoid this trap, the process you are in now starts with the overarching, general, big-picture vision for the farm, then filters ideas and eventual goals through that lens. The result might be the exact same outcome as the example above, but this time it will be based on your bedrock values, vision, and mission. That is the key difference, and it cannot be understated—it is the difference between simply saying what you do instead of what and why or how you do it.

Every business needs a mission statement: one to five sentences that describe your business' overarching purpose. Your mission statement should encompass your values and your vision and can also touch on more concrete topics like your products, market, and management. It should also indicate where your business is headed. Shorter and more general is better. Think of this as the 30,000-foot view from above your farm. Here are some tips for writing a mission statement:

- Keep it short
- Communicate your values
- State your farm's purpose
- Do look into the future
- Don't make it too limiting

If you need an example to get started, here is the mission statement that I created for my farm:

> My mission is to create a farm that honors God through the stewardship of His land, values humane treatment of the animals entrusted to my care, pro-

vides clean, healthy food for my family and others, and develops a spirit of independence and self-sufficiency within myself and my children.

Notice how my mission, in one succinct statement, included:

- The values that motivated me to farm (honoring God through the stewardship of his land)
- The values that determined how I farm (humane treatment of animals)
- The outcomes of my efforts (both healthy food and a spirit of self-sufficiency)

Now it's your turn!

Setting Goals

Goal setting is one of the hallmarks of success in humanity. Top-level athletes, successful businesspeople, and profitable farmers all have one thing in common: they set and strive towards goals! Setting goals focuses your attention and effort, increases motivation, and helps to organize time and resources. It also creates a measurable timeline that allows you to quantify progress, increasing self-confidence and personal momentum. But have you ever stopped to think about what exactly a "goal" is? I mean, a really good goal? Of course, a goal contains the target that you want to achieve, but what else should it include and what shouldn't it? One measure of success in writing a good goal is to use the time-tested acronym SMART. According to this maxim, goals need to be:

Specific
Measurable
Attainable
Relevant
Time-Bound

For our purposes, the goals that you are about to write need to reflect things like what you would like to produce, who will be involved, and what income needs to be earned from the business. An example of a SMART goal might be "Generate enough profit from the farm to cover expenses by the end of the third year." It's specific to a level of profit, measurable via a balance sheet, attainable, relevant to your business, and time-bound—perfect! Notice that it did not include anything like "by selling retail cuts of meat directly to consumers." That is one way you might generate profit but is certainly not the only way. Your goals should not include the "how." That will come later.

Writing SMART goals is easy once you get the hang of it. Here's an example:

Poorly written: I will save money for new equipment.

This is relevant and attainable, but it is not specific in how much money and is therefore hard to measure. It also isn't time-bound.

SMART: I will save $2,100 in the next 12 months so that I can purchase a cattle oiler.

This is specific to what equipment and how much needs to be saved, and therefore it is measurable. It is also relevant and provides a time frame.

If you are anything like me, your brain has already started to categorize and organize this goal-setting process. You could adopt the same categories that I suggested to break apart the values exercise: personal, economic, environmental, and community. Alternatively, you could use the sections of a typical business plan to organize your goals: marketing, operations, human resources, and finances. Both of these would be perfectly acceptable. However, for most of us, the biggest challenge in setting SMART goals is the time-bound part. It is easy to throw out ideas without any time constraints; the real challenge is in determining an achievable time frame to accomplish said goals. Because time frames often provide this challenge, I suggest organizing your goals into short term (1-5 years), intermediate (5-10 years), and long term (10+ years). You'll want to work through which goals you plan to achieve in each time frame, and, by doing so, all of the previously mentioned categories should be adequately represented.

Setting Goals Driven by Mission

Here are a few reminders as you work through each goal:

- Ask yourself whether that goal is reasonably attainable in that time frame.
- Remember you don't need to include how you will attain the goal.
- Make sure you have a way to measure each goal—you can't define success if you can't measure it!

Just like you shared your individual vision with the team that surrounds you, you'll also want your team to identify their own goals for the operation, and afterwards come together to share them. When this happens, for perhaps the first time, real differences are going to appear and will need to be addressed. It is unrealistic to say that you can't move forward until everyone's goals match exactly, but on the other hand they can't be so different that there are no commonalities represented. Now is the time to get everything out in the open and resolve the differences, so that you can move forward united under a common (or at least copacetic) set of goals.

These goal-setting efforts will almost certainly result in a list that is long and perhaps overwhelming. Before we leave this chapter and dive into specific options for branding, marketplace opportunities, and customer communication, you'll need to prioritize your individual and group goals. This will help you focus your resources and time on the most important objectives, allowing you to make strong, targeted management decisions for your business. Keep in mind that you aren't tossing out the remaining goals; you're simply tabling them for future consideration, prioritization and action. That also implies the truth that this entire process is dynamic instead of static, and needs to be revisited early and often as you begin to build your farm into a reality. Ideally, you and your team can use a prioritization process that identifies the five goals that are most important to your business.

Get Strategic, Just Not Yet

Ultimately, you will need to coalesce all of these ideals into a clearly communicated and laid-out strategy that will both serve as a roadmap for turning your dreams into reality, as well as a way to communicate and convince others to help you along the way. Before you can really build a strategic plan, there are lots of things to think through. The next several chapters will highlight many of the crossroads that I faced as I built my farm, including the options I considered and my final decisions. That isn't to say that these need to be your decisions; instead they are intended to provide as much information for you to consider before making your own choices. I recognize that you almost certainly have a set idea already formulated when it comes to many these decisions, so my challenge to you is to open your mind to alternatives and options. Consider all the avenues. Weigh the possibilities. Then pick the one that best fits your values and vision, and build your dream farm from that solid foundation.

Write Your Own Story

Answer the following questions:

What values do you want your farm to be known for?

What is your farm's purpose?

What are the outcomes of your farm and your work?

Based on your answers to these questions, write a concise mission statement that communicates the purpose of your farm and what values you want to be known for. Then write out some SMART goals to help you fulfill that mission statement, keeping in mind that it may help to organize them into short term (1-5 years), intermediate (6-10 years) and long term (10+ years) categories.

CHAPTER 4

The Face of Your Farm

Your brand is sacred. Not just important. Not just a high priority. Sacred, meaning "dedicated to a religious purpose and so deserving veneration." Let that sink in for a minute.

Your farm's brand is a collection of ideas and images, representing the core of who you are and what you want to achieve. It is the customer-facing manifestation of the vision and mission that we discussed in Chapter 3. It represents a sort of business savvy that intentionally develops an image, message, and target audience, and it works to maintain a certain reputation within that audience and must be crafted carefully and protected sacrificially. There is no other piece of your farm that is more important, and every other part, operation, and decision must support and serve—dare I say venerate—your brand. This truth is reflected in the fact that there are literally thousands of brand-management companies all vying for the right to create, share, and protect your brand.

Since I'm the DIY type, we're not going to discuss hiring those companies. Instead, I'll share the steps I went through to create my farm's brand, starting with the name, logo, and initial web presence.

What's in a Name?

For me, choosing my farm name took on the same level of importance as naming my children—or at least I found myself doing the same types of things. I tried to think of all the ways the name could be pronounced (or mispronounced). I considered what the initials would be and how easily they could be turned into something dirty by a school bully. I searched for ways to adequately and accurately convey the meaning, vision, and hopes I had for my farm, and in the end I was very pleased to have arrived at "Pastured Providence Farmstead."

It's worth mentioning that this decision didn't come easy; in fact, it took months of discussion and brainstorming between me and my wife, and the small team of friends and family that we had gathered around us at the time. We had a general agreed-upon framework, but, as usual, the devil was in the details. We knew we wanted to label ourselves a "Farmstead," which is a combination of the words "Farm" and "Homestead." In our eyes, the farm would be grounded in a desire to provide as much as possible for ourselves, become more self-sufficient, and live off the land...all very homestead-type principles. Beyond that, all options were on the table. Here is a partial list of farm names we came up with:

- Grass Fed Farmstead
- Covenant Farmstead
- P.R.A.I.S.E. Farmstead (People Raising Animals in Sync w/ Environment)
- Honey Meadow Farmstead
- Honey Hill Farmstead
- Canaan Acres Farmstead
- Crossing Jordan Farmstead

- River's Edge Farmstead
- Tranquil Acres Farmstead
- Peaceful Pastures Farmstead
- Providence Farmstead
- Peaceful Providence Farmstead
- Providence Pastures Farmstead

As you can see, we covered a fair amount of territory but began to hone in on what we really liked at the end. Providence means "a manifestation of divine care or direction," and spoke directly to the source of our strength, hope, and faith. For a while there, we thought we had arrived at Providence Pastures, but we really liked the switch in order that seemed to put more emphasis on the "Providence" and made "Pastured" a descriptive word, describing how this particular instance of God's providence was carrying itself out in our lives.

Whether you choose a family name, adopt a local landmark, opt to highlight a landscape feature, or are drawn toward a descriptive name that holds meaning for you, the takeaway here is to take your time. Say the name over and over, and don't settle on a decision until an option awakens that excitement inside you that happens when you just know. You won't escape all the ways that someone can mistake your name; in fact I still had at least several people every year tell me that they couldn't find my farm online because they were searching for "Pasteurized Providence." Seriously? Don't worry about those folks—for some, even the lowest common denominator is still too complex. Choose a name that is meaningful to you and your team, that speaks to you and for your vision, and that excites you about the future farm you are starting to build.

Slogans, Taglines, and Logos

Now that you have a long list of words that you just culled through to create your farm name, let's put them to good use—it's time to develop your tagline! A tagline is a phrase that is crafted to represent or reflect your business. Think of it like a motto, or an uber-condensed version of your values statement. Technically, a slogan is different, in that a tagline should be used consistently for the whole company while a slogan is part of a specific marketing effort or individual product. Keep in mind, though, that those definitions were created for multi-billion-dollar, multi-national, multi-enterprise companies with lots of irons in the fire. I don't know of very many farms that have a tagline and also slogans for their specific products, so for the purposes of our conversation we can safely use the terms interchangeably.

My farm's tagline was "Partnering with Creation to Produce Healthy Food." Pretty good right? Well, you wouldn't believe the amount of thought and conversation that went into creating that phrase. It almost seemed harder to condense all of my vision, values, and desires into seven words than it did to write my entire values statement. My goal was to capture the concepts of stewardship, animal husbandry, spiritual guidance, and my effort to provide clean, healthy, unadulterated food for my family, friends, and community.

Obviously, all of that was too wordy, and so the word crafting began. Was I partnering, cooperating, assisting, being guided by, or listening to the Creator to run this farm? In fact, was it with the Creator or Creation where that interaction was taking place, or both? Was I creating the food, or was I just care-taking those that were? Did that make me a producer, a manager, a partner, a farmer, a seller, or a hired hand? Lastly, since I had to pick one word to describe the food that would be created, what should it be? You get the picture: it is a lot of work to create a statement that adequately, completely, and concisely captures your farming enterprise!

Now that you have your farm name and tagline completed, it is time to turn your attention toward the most visible portion of your farm's branding: a logo.

Your logo will represent the "face" of your farm—at least when you aren't there to be the actual face of your farm—so it is important to support your farm's brand with a strong logo. There are actually seven different types of logos. "Lettermark" and "Wordmark" are similar, in that they simply use either letters from, or the full name of, the business. Think CNN, HBO, or NASA for lettermark logos. These all have something in common—they are acronyms for much longer company names. Visa or Google are some common examples of wordmark logos, which are quite simply the company name in a distinct typography. For both these types of logos, the font you choose will help define your brand—you'll want to choose something that is both easy to read and that reflects your operation. You may want to use a wordmark or lettermark logo if you have a memorable farm name or perhaps if you have a farm name that is just really long!

"Pictorial" and "Abstract" are the next grouping of logos, where a graphic alone is chosen to represent the brand. The apple image for Apple and the bullseye for Target are examples of pictorial logos.

Meanwhile, abstract logos use the same concept of an image for a logo but take a more, well, abstract approach. The Nike "swoosh," which implies movement, is a good example of an abstract logo. These types of logos are best used when a business can create widespread brand recognition without having to explain what or who they are. The risk in using this style of logo is that you are unable to explain what your brand is to the consumer, especially at the beginning.

The fifth type of logo is the "Mascot" logo. Kentucky Fried Chicken has the Colonel and Planters Nuts has Mr. Peanut (or did until his untimely death in 2019). In the farming arena, I don't recommend making a cute little lamb as your mascot, lest you spend the remainder of your days explaining to the vegans why you saw fit to chop "Lester the Lamb" up and turn him into leg of lamb! Unless there is something so special about your locale, product line or farm's personality to warrant it, I recommend staying away from mascot logos completely.

"Combination" logos do just exactly what it sounds like: combining elements of lettermark/wordmark and pictorial/abstract/mascot. The benefit of this type of logo is its versatility in using different elements to communicate and enforce your company name and branding.

Wrapping up the list is the "Emblem" logo, where letters or words are used within a symbol to create a badge, seal, or crest. Harley-Davidson motorcycles and Starbucks Coffee would be examples of an emblem logo. A challenge with these logos can be the detail involved, depending on how the logos are used. Too much detail can be very difficult to replicate in different formats, such as embroidered shirts, business cards, letterhead, etc.

The logo that I chose for Pastured Providence Farmstead falls into the Emblem category and, along with combination logos, is where I would recommend that beginning farms concentrate their design efforts. They lend themselves to start-up operations and are the easiest to begin telling your story through. So how did I muddle through all of that information to land on my farm's logo? To start, I had a firm grasp of my farm's mission and goals. Drawing from them, I developed a vision of how I wanted to convey that with imagery: a multi-species pastoral scene incorporating my tagline (more on that in a minute) and a bible verse that represented my farm's "life verse."

For your own logo, start by asking yourself what values and ideas you want to capture in your logo. Then make a list of what images, colors, and words might help communicate those values. This should lead you to a basic idea of what you might like in a logo.

Much like the process you used to create goals and mission statements, there are some considerations you'll want to take into account as you conceptualize your logo—primarily having to do with limitations. Again, since your logo is the face of your operation, you'll want to make sure that whatever you ultimately decide does not present any limitations in what it conveys and how it can be used. *Representative* is good.

As you choose imagery, fonts, and words, be sure that you aren't so detailed that it narrows people's perceptions of what your farm represents. *Simple* is good.

As you start to consider the elements that you want to include in your logo, think about the inherent design limitations and how you anticipate it being used. For example, one of the downfalls of an emblematic logo like mine is the lack of flexibility, something that I definitely found limiting when printing small items like business cards and marketing brochures, as well as embroidering shirts. It got to the point that you almost couldn't read some of the text on my printed materials, and the logo had to be ridiculously large on a shirt to be able to read properly! If I had to do it all over again, I would choose less fine detail. *Adaptable* is good.

Finally, multi-color is great right up until you have to pay a printer triple the cost for all those colors on a banner or a t-shirt! Consider a monochromatic logo (or at least a design that can be printed in monochrome if need be), using black & white as the other two colors. This will keep your costs low and give you lots of options for black & white printing or inverting the colors. *Recognizable* is good.

With these considerations in mind, let's look at a few examples.

First, a bad example: a rejected design for my farm.

It is multi-color, which is OK, but all the colors reflect a sickly shade of green, while the tagline and bible verse text are too small to even be read at t-shirt or website sizing. It has creepy hands that don't even look like hands, which were intended to symbolize holding the land but instead ended up detached and randomly floating outside the core logo. Lastly, the trees have an enormous amount of detail, yet the fence looks like my four-year-old drew it. Not an awesome "face" to my farm and definitely not conveying the values and quality I had hoped to, which is why I went with a different design!

Contrast those problems with Joel Salatin's logo for his Polyface Farm: his logo is a single-color sketch of a tree. The stylized branches nested within are drawn in the shapes of a cow, chicken, and fish. Represented there is Joel's connection to the land and roots firm on his ground, the multiple species of animals literally reflecting his tagline, "The Farm of Many Faces." *Simple, adaptable, recognizable, and representative*—all the elements of a good logo.

A question that you'll need to tackle somewhere within this process is whether to, and how to, integrate your tagline with your logo. In my case, I chose to include it within the logo itself, arching over the pastoral scene in the space that created the circular emblem. Alternatively, Polyface Farms chose to keep their logo separate from their farm name and tagline, "The Farm of Many Faces," so that now they can use them together when appropriate but also use the logo independently if needed. Keep in mind as you think through design options that less text within your logo will yield greater flexibility in terms of sizing changes.

Not all of us have the talent or capabilities to design our own logo. I know I certainly didn't! There are logo designers everywhere that will work with you to create an amazing logo, but beware that they will likely cost you a pretty penny. If you want to ride the line between cost and production, then you might consider Hatchwise, an online design competition where you state your logo goals and restrictions and designers from all over the world compete to create the best logo, ultimately winning the amount that you offered as a prize. The more money you offer, the more likely you will draw quality designers vying for the win. I had over 20 submissions for my logo design contest, in which I offered $229. Some were clearly just thrown together and terrible, but there were two front-runners who also made some adjustments to their original designs throughout the process. After selecting the winner, I received a full digital pack of logos, including multiple file formats and a black-and-white option. If you do decide to use a designer, you can expect to spend at least $400 to $700 for a final design (with some edits) and all final design files.

There are many things to consider when it comes to your logo, but I can't stress enough the importance of getting it right. Once you have this customer-facing surrogate in place and have built your brand around it, the last thing you want to try and do is alter or adjust it! Spend the time now to get this aspect right the first time. Create a list of words that reflect your values, desires and vision, likely starting with the words that were in the final running for your farm name. Use a thesaurus to find alternative words that have the right connotations as well as the "feel" that you are trying to communicate. Take an initial stab at writing your tagline, or even a short list of alternative options, but then remember to step away for a day or two to let things ruminate before deciding. Consider the different types of logos that might fit your needs, and the inherent limitations within all of them, then either hire a professional design company or create a design competition with Hatchwise. And as always, take your time, involve your team, and enjoy the process!

Trademarks, Copyrights

As I was working through creating my logo and tagline, the question of "legality" popped up time after time. This was especially true since I was using the pay-to-play services of a random artist through the Hatchwise website. I can't tell you how many times it came up in conversation: "What do we have to do to legally use the new logo?" It's a potentially confusing topic that seems like a waste of time when all you want to do is farm, but we'll cut through the fog and create some clarity about what trademarks are and if they are necessary.

Let's start with some definitions: a trademark is a distinctive word, name, phrase, symbol, design, or other device used by a company or person to distinguish its products or services from the products or services of other companies. More than a simple description, according to the US Patent and Trademark Office it "identifies and distinguishes the source of the goods of one party from those of others," and as such is protected by law. A trademark is different than a copyright, which is a form of protection provided to the authors of original works of authorship, including literary, dramatic, musical, artistic, and other intellectual works. So you would potentially want to trademark your logo and copyright your blog content, for example.

Existing trademarks can be searched on the TESS (Trademark Electronic Search System), which incidentally is a great place to start when you are researching your logo. It would be unfortunate if you did all the work or paid an artist for a design, only to find that it was too similar to an existing trademarked image. You really have to hone your search skills, though. When I searched for "farm" it came up with 38,200 records. Then when I tried "pastoral scene on a ribbon background" I got 871,972 records...great. A trademark lawyer would probably pay for themselves in this part of the process alone!

Once you have a logo, you have to establish a "basis" for requesting trademark protection, meaning you have to either demonstrate its use in commerce, or at least declare an intent to use it in commerce. That's right—you actually have to start using the logo on your products as part of the trademark approval process. Backwards I know, but you send a picture of you using your unprotected logo in commerce (meaning you've already spent the money to print labels, embroider shirts, or more), along with a textual description of your logo and a hefty fee of $375 per class to the government.

Then the Patent and Trademark Office will (hopefully) send you a certificate saying that the logo is yours and only yours to use.

This process already didn't make sense to me and obviously wasn't going to keep me from using my logo, but the final kicker for me was the "per class" requirement. In their infinite wisdom and greed, our fine bureaucracy requires a separate application, and fee, for each class of goods. So if you wanted to use your logo on packages of meat (class 29), honey jars (class 30), and screen-print it on a shirt (class 25), then you would have to trademark all three uses separately, to the tune of $1,125. Heaven forbid if you had the great idea to put the logo on a vehicle or trailer (class 12), create letterhead for the farm (class 16) or advertise for school field trips (class 41)!

Needless to say, I skipped the trademark step. If someone liked me enough to copy my logo, then my plan was to try and take it as a compliment. And if I got successful enough someday that someone might actually care to copy me, then hopefully I'd be financially solvent enough to afford protecting my logo through trademarks. Not having your logo trademarked doesn't keep you from using it however you want, as long as it doesn't interfere with someone else's branding, so being courteous and mindful goes a long way toward avoiding a lawsuit! Obviously, I can't officially tell you to ignore trademarking your logo, so I will simply leave it at this: I farmed successfully for years (gasp!) without a trademark. If you decide to register for a trademark, more information on the details can be found here: *uspto.gov*.

Copyrighting is an equally muddled topic, but at least here you have some concrete actions you can take to put the protections of copyright into effect. As previously discussed, copyrights are intended to protect original works of authorship, including literature, art, drama, and music. Copyrighting gives the author the exclusive right to reproduce, distribute, perform, or display said works, or authorize others to do so. If you decide to start a blog, write a book, or jot down the lyrics to a song that came to you while you were rolling around your land on the tractor, then you might consider copyrighting that original work of art or content. The interesting thing about a copyright is that it enforces a right that exists regardless—in other words, you don't actually need to register a copyright for it to be yours.

Essentially, as soon as you create any given work, your rights are automatically secured without any action on your part. All you are doing when you request your copyright is notifying the government that you have created a work and it needs to be protected (think "register" instead of "request"). It is a legal formality that makes the creation of your work a matter of public record. There is no requirement to register your work, and you are protected with or without registration. That being said, there are a several incentives to registration: first and foremost, you can't file an infringement lawsuit without registering the copyrights. In addition, registration establishes evidence supporting the validity of the copyright and makes certain compensation for attorney's fees available to the owner.

Registering a copyright involves two of our government's favorite things: paperwork and fees. You can file the request online or via hardcopy (some specific works require hardcopy filing), and the fees range anywhere from $35 to $220 depending on the type of work. The final requirement is a non-returnable "deposit" (i.e.: copy) of the work that gets filed with the Copyright Office and eventually makes its way to the Library of Congress. Efficient...I know. After all this is submitted, you would hope to receive your certificate of registration in the mail, or a denial letter if you forgot to dot an "i." If you are denied, you can submit for reconsideration, for an additional fee, of course!

Once I figured out that copyright protection existed without me doing anything special, and that registering the official copyright was such a pain in the you-know-what, I again decided against officially registering any of my fine literary work. At the time, the only thing that might have been even close to applicable was the content I was posting on my blog, which we will discuss more in Chapter 8. Since the blog was housed within my website, I did add the copyright notice for "visually perceptible copies" to the bottom of all of my webpages, in order to notify the public that the content contained therein is protected by copyright laws. Basically, that consists of the copyright verbiage or symbol (©), the year of publication, and the name of the owner. So mine reads "Copyright © 2013, Pastured Providence Farmstead."

The lesson here is really this: trademarks and copyrighting exist and they are technically applicable to a new farm business that is creating logos, taglines, and intellectual content, but ultimately it all falls into a category that I love to hate—government bureaucracy that creates work simply for the sake of work. I'm certain that my publishers are rolling their eyes, but this is their world and ours is out on the land. So please do your due diligence, be aware of the information, put whatever level of protection around your branding that you feel necessary, but don't let it keep you from taking your next steps. On the very off chance you would like more information than I have provided here, you can visit *copyright.gov* for all the gory details.

Web Presence

After deciding on a farm name, tagline, and logo, the final step in branding your farm is securing your "web presence." Chapter 8 is entirely dedicated to ideas on how to build your customer base, including online options and opportunities. For now you just need to secure the foundation for your eventual virtual presence, in the form of domain registration. There might be some of you who are thinking to yourself: "I don't need a website, I'm just going to sell all of my awesome food to folks the old-fashioned way." To be blunt, you are wrong. Whether you think you need one or not, go ahead and work through this process and purchase some domains. I believe that you'll be glad you did in the end.

All domain names are registered through a non-profit organization called the Internet Corporation for Assigned Names and Numbers, or ICANN. Can you imagine the water cooler conversations that go on in that office space? As expected, their website is pretty sweet...you can search existing domain names with their online Lookup Tool (*lookup.icann.org*) and register your unique domain name directly with them or use a website-hosting or domain-registrar service like GoDaddy, Squarespace, and others. ICANN sets the cost of domain registration, so you don't have to worry about one company charging a premium for your desired domain, but sometimes they offer package deals to get a beginning website or blog up and running in conjunction with your domain purchase; so if you already have an idea for who you want to host your website, etc., then feel free to set up an account with them.

It might seem a little like the virtual cart-before-the-horse, but the reason I am tackling domains now is because it needs to flow from your farm name. In an outlandish example, you would never choose to name your farm "Pastured Providence Farmstead" and then have your website domain be *farmsteadinthepasture.com* or something else equally unrelated! You don't have to build your website yet—we'll get to that eventually—so instead think of a domain as the web address to where you will eventually send someone to view your website. Your domain is the online version of your logo, in that it serves as a sort of first impression for those just meeting your farm online, so you'll want to make a lasting, positive impression. Some rules of thumb to keep in mind are:

Stick with a .com or .org, as these are most popular and easiest for users to remember. There are a multitude of alternatives, even including a .farm domain, but if you can find a suitable domain with a .com then you will be much better off. This isn't the time to get clever!

Your domain should align closely with your farm name, if not BE your farm name. For my farm, Pastured Providence Farmstead, I chose the domain *pasturedprovidence.com*.

The shorter, the better. This is to keep it simple for folks to remember, but also because shorter URLs are easier to fit onto things like business cards, signage and mobile phone screens.

Avoid hyphens and numbers (unless that is specifically part of your brand).

Consider using strong keywords that reflect your business—this can help people find your farm easier as it improves your SEO (search engine optimization) ranking.

If you get stuck trying to come up with your domain, try using a domain name generator. These are web portals that can take your ideas that don't quite work and generate alternatives that might! Some platforms that offer this include Wordoid, Lean Domain Search, and DomainHole.

To make things even more interesting, you can buy and "point" multiple domain names towards your primary domain name, and so I recommend purchasing a suite of domain names that are similar to the one you originally chose. Why? Because humans are fallible, domains are cheap, and you need to do everything you can to mitigate some of the human factors that might prevent someone from finding your website by searching incorrect terms. In my case, I spend $109 per year to maintain the following domains:

- *pasturedprovidence.com*
- *pastured-providence.com*
- *pasturedprovidencefarm.com*
- *pasturedprovidencefarmstead.com*
- *providencepasture.com*
- *pastureprovidence.com*

When I first started the farm, I think I had twelve domains total, all of which re-directed to the *primarypasturedprovidence.com* web content. Why not? We'll discuss how to build the content of your website later on, but as part of your initial branding efforts (whether you think you'll ever need a website or not), go do some domain research and buy yourself a primary domain along with as many combinations or incorrect iterations that you can think of and afford. Your brand thanks you in advance.

For those who are wondering, I never did go back and purchase *pasteurizedprovidence.com*...I preferred to let those folks pass right on by!

Write Your Own Story

Write a list of possible farm names, then pass them through the "bully" filter to eliminate ones that could be misinterpreted or mistaken for something else. Which words adequately and accurately convey the meaning, vision, and hopes you hold for your farm? How do these key words and concepts influence or affect possible logo designs? Feel free to doodle or sketch a few initial ideas.

CHAPTER 5

Business Structures and Tax Implications

DISCLAIMER, PLEASE READ: Portions of this chapter were written by an attorney holding an active license to practice law in the State of Ohio. The excerpted content is based solely that attorney's knowledge of Ohio law. Any legal advice provided in this lesson is for educational purposes and should not be construed as providing legal advice to non-Ohio residents. Readers from states other than Ohio should consult a licensed attorney in their state of residency or operation.

How is that for a start to this chapter? You know what they say: Free advice is worth what you paid for it. At least you paid for this book, so you can feel good about the advice contained within these pages. However, when I began to discuss my intention to provide legal advice in this chapter, several friends stepped up and helped me understand that that might not be the best option. In fact, it might have actually been illegal. To keep myself out of hot water and on the right side of the law, I enlisted the assistance of my good friend Ryan Conklin. Let me tell you a little more about him.

Back in 2016, I was honored to be accepted into the Ohio Farm Bureau Federation's leadership development program called AgriPOWER. Twenty-five people from all over the state, representing a diverse cross-section of agricultural professions and interests, met together over the course of nine months to be immersed in all things farming. I started the program as an outlier given my...unique... perspective on growing and raising food, but was welcomed by my classmates nonetheless. Over the course of my time in AgriPOWER I explained (and sometimes defended) regenerative agricultural principles, listened to the views of those who would be considered "conventional," and gained an entirely fresh recognition that my so-called conventional classmates and I had a whole lot more in common than we had differences. Did we always see eye-to-eye? Of course not. But I was grateful to recognize that their hearts were sincere, their concerns genuine, and their intentions honest. To my shame, that is not what I expected to find within this group of supposed adversarial outlooks, but I'm glad to say that I was dead wrong. Several members of that group remain my close friends to this day, and Ryan Conklin is one of them.

Ryan comes from a farm family and is a licensed attorney with Wright & Moore Law Co., LPA, in Delaware, Ohio. He is a good man, extremely smart, handsome to boot, and his expertise was invaluable to me in writing this chapter. The excerpted text are his words, partly because he knows exponentially more about the subject than I do and partly because I can't legally give legal advice, hence the disclaimer.

Do You REALLY Need a Lawyer?

I've mentioned this several times already, but I have a disease. It's called "Do It Yourself" disease, and it is both a blessing and a curse. I hate to pay someone for something that I think I can do myself. Often this has saved me money and resulted in generally positive outcomes. One glaring exception to that truth revolves around the content of this chapter. My story and the decisions I made in this arena should be regarded as cautionary tales and bad examples. I did not do this part of building my new farm correctly, and as such I put my farm, finances, and future at risk.

I decided on an LLC structure for my farm, mostly because that's what it seemed like everyone else was doing. It was also was a risk management decision intended to shield my personal assets from any potential lawsuits. Within such a litigious society as ours, raising food alone is opening yourself up to serious scrutiny. Add in farm tours, events, labor provided by friends and family, and the dangerous nature of farming in general, and all of a sudden you have an extreme level of risk compared to most businesses. So after we finally settled on Ohio as our location (business structures are regulated at the state level), I did what I always do: I started a Google search, and that is where my mistakes began.

Mistake #1 was when I found a free website that would take my input and turn it into an "operating agreement," complete with all the things I legally needed to include. This seven-page document included the LLC's name, named me as the registered agent, established our initial financial contributions from the two members (Heather and me), and detailed the interest in the company (50 percent each). After that, it contained all the legal mumbo jumbo about powers of legal representation, dissolution, liquidation, and settling disputes. But who really cares about any of that because it's not going to happen right?!

Yeah...

The operating agreement was signed and mailed it to the State of Ohio, along with a check for the filing fee of course, and just like that Pastured Providence Farmstead, LLC was born on October 6, 2013! Part of the agreement stated that the LLC members would meet quarterly and minutes would be taken, in order to preserve the integrity of a separate business entity different than my personal financial interests. We met in October 2014 and January 2015. That's it. After that, we didn't make it a priority and "real" farming got in the way—Mistake #2. By not honoring the requirement to meet and document, we were already setting the stage for the newly created LLC to be considered a sham if we were ever sued or held liable for damages.

Even when it came to ownership, I never really utilized the LLC like I should have...Mistakes #3 through 69. This can be partly attributed to the murky nature of running a business on the same land where I lived, but that is still no excuse for taking the shortcuts I did. If I was to truly do things right, Pastured Providence should have paid Paul Dorrance rent for use of the land that was used. Pastured Providence should have purchased the tractor and hay equipment instead of Paul Dorrance. Even the insurance policy for Pastured Providence's farm product liability got lumped in with Paul Dorrance's property and vehicle insurance because, you guessed it, there was a price break for multiple coverages and I was too cheap to pay extra for the stand-alone farm policy. You get the picture. Mistake after mistake on my part.

So what would have happened if someone was injured on my property by something that I should have seen and corrected? Would my LLC have protected my personal assets from litigation like I intended it to? Absolutely not. I think it is fair to say that I made every mistake you could make in my journey. Thank God nothing bad happened because of those mistakes, but that doesn't make the risk worth it. I never really questioned the concept that finding a good lawyer would keep me from making the mistakes that I did, but I struggled mightily to justify the initial expense for something that I neither understood nor valued. I cannot recommend that path for you. Take my advice…I mean, Ryan's advice… and find a good agricultural lawyer to assist you with setting up your farm's business entity. Your future success deserves to be set on a solid structural foundation. Speaking of, here is what Ryan has to say about finding and selecting an attorney:

> There are a host of factors to consider when hiring a legal counselor. Geography, workload, availability, cost, firm culture, and technical knowledge are among the variables that come into play. To help with your search, here some key points to consider:
>
> The most important factor is actually a concept. At the end of the day, you're looking for an attorney that fits your goals. Take the following into account:
>
> - Is the attorney familiar with farm knowledge, finances, taxes, and practices?
> - Is the attorney able to complete the requested tasks within your timeframe?
> - Is the attorney making recommendations based on your goals? Or telling you what to do without asking about your goals?

- Do you feel comfortable and relaxed when talking with your counselor instead of stressed?
- Have you spoken with other farmers or families in the area about their experience with a particular attorney?
- Have you watched the attorney present on important farm-related subjects or read farm-law articles drafted by the attorney?
- Does a practitioner or a firm have a succession plan? Or will you need to change attorneys again within a few years?
- Does the attorney regularly communicate with you regarding pending tasks?

An attorney does not need to check off all these boxes. Taking care of most of them will be a good sign. Additionally, you can weigh the factors based on which ones are most important to your farm and family. In the end, determining your "fit" with an attorney is a subjective test and can vary substantially.

Geography should be a factor in your search. In order to nail down an attorney who is a fit for your operation, you may need to travel. While there are surely attorneys in your town or county, they might not be qualified agricultural practitioners. Be willing to explore your options around your state. Some clients even prefer the privacy that comes with an attorney located in another town or county. Take advantage of video conferencing options when available to save time and resources on long trips.

When consulting an attorney for the first time, check if he/she is willing to give you a free consultation. Obtaining a consultation upfront gives you a chance to screen the attorney to see if it is a good fit, and it mitigates your financial risk.

How do I change attorneys?

Attorneys are bound by certain ethical standards of practice that govern our providing of legal advice. One of those standards states that the client controls the representation. Furthermore, the attorney is only a caretaker of the client's file. The ownership of that file remains with the client the whole time.

What does this mean? It means that you can change attorneys at any time by requesting the delivery of your file. An attorney is not permitted to hold your file hostage in order to keep your business. You may need to change attorneys if your current counselor retires or cannot provide the advice that you need. The attorney-client relationship is not a marriage and therefore can be concluded at any time.

Are there organizations that can help me find a reputable farm attorney?

Yes, in fact there are three great choices for your search. First, your state's cooperative extension may have a farm law specialist who regularly works alongside private practitioners. He/she may maintain a preferred list of attorneys to help with farm matters. If your state extension service does not have a farm law specialist, your county agent or a farm management expert may know some preferred attorneys.

Similarly, your state or county farm bureau office may have screened some attorneys through meetings. Farm lawyers should regularly be in touch with farm bureau offices about presentations, policy, advocacy, changes in the law, or other matters.

Lastly, a state or national trade association is a great option. The American Agricultural Law Association is the national trade group for farm lawyers and can assist with your search. Likewise, if your state bar association or bar group maintains an agricultural law committee or section, that may be another source of potential attorney options.

Are some organizations bad options for your search?

I would recommend avoiding online legal headhunters or other online legal services platforms. Reason being, my experience suggests these services result in pairing farmers with general practitioner lawyers who might not possess the requisite farm knowledge. Furthermore, these services may provide cookie-cutter legal services for all clients. This means the solutions vary very little from client to client. Your farm has unique needs, and your legal solutions need to match those unique needs. Agricultural lawyers are a special subset of practitioners with unique knowledge that other lawyers do not possess. Many come from farming backgrounds, assist other farm clients, and have catered their education and practice around agriculture. When hiring an attorney to help your farm, consider the special needs of your operation and the need to consult a special counselor.

Choosing a Legal Structure

Creating a business through which to operate your farm is called "incorporation," regardless of the entity structure you choose. However, it is worth noting that you don't have to incorporate in order to farm successfully. Many farms operate unincorporated, without a separate business entity in place. The choice to incorporate is based on your needs, desire for risk management, and requirements for legal separation of personal and business assets. In Ryan's words:

> Whether you should incorporate will depend on one thing: your goals. Here are some examples of goals that some farmers use to justify incorporation:
>
> - **Protecting personal assets:** this goal is intentionally at the top of this list. It is the most common reason farmers form a business. They are seeking to protect their personal assets (house, bank accounts, retirement, etc.) from accidents that occur on the farm.
> - **Formalize management:** occasionally companies will be used by farmers to designate official management roles. In any small business, identifying managers is a key step. For a farm, it can assist with workflow management, task allocation, and decision making.
> - **Assist with succession planning:** businesses are very common succession planning tools and can be used to bring the next generation into the farm. After all, buying 1/10th of a company is easy and is just paperwork, but buying a 1/10th interest in a farm is tougher.

- **Clarify tax circumstances and record keeping:** this goal boils down to two questions: 1. Who gets to claim deductions? 2. How is income divided amongst farm participants? Instead of families fighting over who gets to use a deduction, or how farm income is split, operating within a company solves both problems.
- **Consolidate family or business operations:** some farms feature multiple family units coming together to raise livestock or grow crops. Each family separately owns its own animals, equipment, inputs, and grain, yet the crops and stock are raised together. A formal business provides a framework for consolidating these businesses into a more efficient system.

This is just a partial list of considerations, but Ryan's point is well taken in that it all goes back to your vision, mission, and goals for your farm. If those goals are leading you toward incorporation as the right answer for you and your situation, then great! However, if you finish reading this chapter and don't find a compelling reason to incorporate, just know that you are in very good company and can feel good about keeping your money in your pocket. To help inform that decision further, Ryan discusses the most common types of incorporated businesses.

Sole Proprietorship

This is the bare bones option, and bare bones would even be considered a stretch. Under a sole proprietorship you are operating the business individually with no formal structure, no distinction between your personal activity and your business activity. Therefore, there is absolutely zero protection for your personal assets if a farm-related accident occurs. State laws surrounding sole proprietorships tend to be very minimal. Individual business owners who have not incorporated are automatically assumed to be operating a sole proprietorship.

Partnerships

The next rung on the company ladder is partnerships. Here, two or more people are engaging in some sort of collective business activity. At least two partners are required for a partnership, otherwise an existing partnership will be dissolved. You may have four options for the type of partnership you want to form: a general partnership, a limited partnership (LP), a limited liability partnership (LLP), or a limited liability limited partnership (LLLP).

All partnerships require the designation of general partners to manage the company. Within a general partnership, any general partner can take any action on behalf of the group and bind everyone to that decision. Although this provides immense flexibility, it also means that each general partner is on the hook for any accidents caused by or debts incurred by the other partners. Limited partners, such as those found in LPs, LLPs, or LLLPs, have more limited authority to bind the company, but in return they get some personal asset protection. For this reason, if a family is considering the formation of a partnership, it will almost always form an LP, LLP, or LLLP in order to achieve some liability shielding.

Traditionally, partnerships were formed to capitalize on tax benefits. Today, the most common reason general partnerships are formed in farm families is for Farm Service Agency payment limitation benefits. However, these payment limitation benefits will strictly apply to large farms. If you are not a large farm, your best options lie elsewhere.

Your state has likely adopted some version of the Revised Uniform Partnership Act to govern partnerships within its territory. Your state partnership act will dictate whether your partnership needs to be registered in order to operate, or if you can operate without notifying the state. If you are operating a partnership and have not created a formal partnership agreement, then the partnership act in your state will fill in the blanks and dictate the rules for your business.

It is crucial to keep in mind that your state law might automatically allow for the creation of a general partnership, even if you take no formal action. For example, Ohio law states that two or more persons who share profits together are automatically considered a general partnership. Therefore, if one partner causes an accident or incurs a debt without any involvement from the second partner, the second partner may be personally liable for the accident or debt. This is part of the reason why correct incorporation is an important step for the protection of personal assets.

Corporation

If you are seeking formalities at the highest level, look no further than the corporation. Corporations have a host of requirements that must be followed in order to operate legitimately. Here are some examples of those formalities:

- Issuance, management, and valuation of company stock
- Holding of annual shareholders meetings
- Keeping minutes of annual shareholders meetings and directors' meetings
- Formal notices, including the timeframe for providing notice and the method
- Election of officers and limitations of duties for those officers

In the end, correctly managing a corporation requires great effort and great discipline. If you are not a fan of paperwork, it might be a good idea to look elsewhere.

So, what is the benefit of corporations? Liability protection. Prior to the creation of the limited liability company (which is coming up next), a corporation was the best way to operate a business while protecting your personal assets from an injured party or a creditor.

Much like partnerships, your state may have adopted some form of the Model Business Corporation Act. Once again, this set of state law will govern many of the areas discussed above, business registration, naming, and other areas. Unlike partnerships, corporations must file their articles of incorporation with the state in order to be recognized. Furthermore, if you amend those articles, the amendment must be filed. However, this does not mean the corporate bylaws and regulations must be filed with your state; those can typically be kept private.

Limited Liability Company (LLC)

LLCs burst onto the business planning scene in the 1990s and have since become the preferred choice for new entities. LLCs give planners immense flexibility in terms of setup, management, structure, and operation. Remember all the formalities for corporations? They are largely eliminated when running an LLC. The operating agreement, which is the governing document for an LLC, is the vehicle for that flexibility.

Like corporations, LLCs provide great liability protection. This protection is for accidents caused by employees and co-owners, and for debts belonging exclusively to the LLC. Keep in mind that if you personally cause an accident while performing LLC functions, you still might be personally liable for injuries or damages.

When forming an LLC, selecting a management structure is an important step. One option, a member-managed LLC, involves the ownership (or membership) of the company managing its day-to-day affairs. The alternative, a manager-managed LLC, comes to fruition when the membership of the company identifies non-ownership parties to handle business efforts. Most farm families elect a member-managed business format since they are so closely involved in the operation.

Again, LLCs in your state may be governed by some variation of the Revised Uniform Limited Liability Act. In order to operate your LLC, registration with your state is likely required. This registration, often called "articles of organization," should not require frequent edits. Like corporations, the operating agreement for your company would not require filing with the state. If you do not complete an operating agreement, your state law governing LLCs will step in to govern the relationship between LLC members. Therefore, if you want to customize your business governance and experience, completing the formation process in its entirety will relieve the need for state law to be involved.

Trade Names

Though it does not qualify as a formal business entity, some business owners will register as a "doing business as," or DBA. For example, your business name might be Smith Farms LLC, but you would like to market your cattle as "Circle S Farms." Rather than create a new company, you can use the trade name "Circle S Farms" as a DBA and operate it through Smith Farms LLC. In this instance, all business activity conducted under the Circle S Farms name is run through Smith Farms LLC. Individual persons may also register DBAs.

Typically, states require the registration of DBAs or trade names in order to be valid. If you are operating as a DBA and another person or entity has already reserved that name, you may need to discontinue using the DBA.

Remember that you don't have to establish a formal business structure in order to successfully farm; plenty of farmers are conducting business as de facto sole proprietors all over the country. However, as our society gets ever more litigious and sue-happy, compounded by the increasing costs of healthcare and already elevated risks of producing food for others, I highly recommend that you consider incorporating your business. Choosing the right structure is key, based on your goals for the farm and guided by a licensed professional in your state.

Which Tax Structure Is Right for Me?

DISCLAIMER, PLEASE READ: Neither Ryan or I are registered tax professionals, nor do we hold any tax-related licenses, certifications, or degrees. The information contained here is based solely on our firsthand knowledge from our educational and professional endeavors, working with clients on tax issues, working with tax professionals on client issues, and research. All business plans should be reviewed with a tax professional prior to finalization.

If you thought a discussion on lawyers was bad enough, why not add another disclaimer to the mix and talk about another topic that farmers avoid like the plague: taxes! You might despise them, but you sure can't ignore them. Taxes play a part in just about every financial decision that you make on your farm. For our purposes we'll concentrate on federal income taxes, as state, employment, and sales taxes are highly varied across the country. For now, just know that your decision regarding your business entity will almost certainly affect your state taxes, but a tax professional will need to provide that information. The following are Ryan's thoughts on the tax implications of each business structure:

Sole Proprietorship

If you recall, a sole proprietorship is a business structure where the owner and business are one. There is no legal distinction between the personal assets of the business owner and the business assets used on the farm.

Taxes for a sole proprietor are simple. Since there is no formal business, there is no tax return to pay for that business. Therefore, on your personal income tax return, you list the profits or losses sustained from your farming enterprise. Typically, this is accomplished on Schedule C or Schedule F of your return.

You cannot make any elections to change this tax status for a sole proprietor. Nor are there any major pros or cons for a sole proprietor's taxes.

Partnership

A partnership, which is a business owned by two or more partners, is a great place to introduce the concept of pass-through taxation.

A pass-through business is one that does not pay any taxes directly to the Internal Revenue Service (IRS). Instead, the taxes owed on any profits made by the business pass through directly to each partner. Partners then list the profits on their 1040 forms and pay taxes on those profits in proportion to ownership. These profits are taxed as ordinary income on your personal returns and the tax rate may change based on the total income reflected on each 1040 filing.

An informal partnership may not require a separate tax ID number. Instead, the partnership would be identified by the social security numbers of a partner(s). However, if you have a formal partnership (registered with the state, many partners, drafted a partnership agreement) it may be best to obtain a separate tax ID number for the business. There are no other tax structure elections you can make under a partnership.

In the early days of farm business planning, partnerships were a popular selection because of the single layer of taxation. As the partnership became obsolete as a business structure due to liability concerns, it remained a viable tax structure to employ in a business.

Corporation

Remember all the legal formalities for corporations from the last lesson? Sadly, it doesn't get any better for corporations on the tax side. When you form a corporation, your personal assets gain liability protection, but it comes at the expense of unfavorable tax scenarios.

When selecting a tax structure for a corporation, you have two choices: a C-corp or an S-corp. Both have their own benefits, but also major drawbacks. These next two charts, both from *PayTech.com*, provide a great explanation of the differences between the two.

Figure 5.1

FEDERAL & STATE DIFFERENCES	
C Corporation	**S Corporation**
• Taxed as a separate entity	• Avoids double-taxation
• Shareholders are not taxed	• Corp. doesn't pay income tax
• Shareholders pay income on payments from the corp.	• Can include losses on personal tax returns
• Is always recognized by the federal & state	• Not recognized in certain states

OWNERSHIP & ACCOUNTING DIFFERENCES	
C Corporation	**S Corporation**
• More tax-free status on fringe benefits	• Less tax-free benefits for shareholders
• Can have multiple classes of stock	• Limited to one class of stock
• Can choose when fiscal year ends.	• Fiscal year must end December 31st
• Bigger corps. required to use accrual accounting method	• Only those with inventory have to use accural accounting method

Figure 5.2

Double taxation is typically a non-starter for many farm businesses. It means that the corporation files its own tax return each year and pays its own taxes. Then, if the corporation distributes dividends to shareholders, the shareholders pay taxes on those distributions. Another negative for both corporate tax structures, is that they are notoriously difficult to dissolve without triggering taxable events.

Remember, the default tax structure for a corporation is a C-corp. If you would like to change to an S-corp, you need to make the election to do so. Changing to an S-corp must be completed on a specific timeline, and you should employ the assistance of a tax professional to assist with filing the paperwork.

Electing a corporate tax status is a very numbers-intensive effort. Even with the tax reform passed in 2017, small, closely held businesses struggle to find justification for using a corporate tax structure. A well-researched decision must be made if you are selecting a corporate tax structure over a pass-through tax structure.

Limited Liability Company (LLC)

When it comes to LLC taxes, all you need to do is read the previous two sections on corporations and partnerships. That's right, the taxes for LLCs are derived from those two structures!

There is no such thing as an "LLC" tax structure. Therefore, when you form an LLC, you get the choice between being taxed as a partnership, C-corp, or S-corp. The same advantages and disadvantages outlined above apply to each structure, except now you are operating within an LLC. A partnership tax structure is the default selection, meaning you'll automatically receive pass-through taxation status. If a corporate tax system is your preferred route, you will need to submit special filings to the IRS very soon after incorporating.

Now, there is one distinction here between multi-member LLCs and single-member LLCs. Multi-member LLCs, or those with two or more members, can make an election between a partnership or corporate tax structure. Those multi-member entities must submit an annual separate tax return like any other business. However, a single-member LLC is recognized as a "disregarded entity" by the IRS. This means it does not need to file a separate tax return. Instead, the company earnings are reported on the sole owner's personal tax return. If a single-member LLC adds new members, it will need to change tax statuses.

A final point about a tax structure election for an LLC: once you make the first election (partnership or corporation) it is very difficult to change statuses. It may involve additional IRS filings, asset changes, regulatory delays for the new tax structure, and other hurdles. Therefore, selecting the proper tax structure upfront is essential.

As you can see, your selection of business structure and its tax implications is absolutely critical to understand. Yes it represents equal parts frustration, convolution, and red tape, but it cannot be ignored. Ryan's words are just the tip of the iceberg on this subject, so after you find yourself a good lawyer, ask them for their recommendation for a good accountant. You'll need both to help you avoid the mistakes I made, keep your business protected, and guard your personal assets.

Operating a Legitimate Business

Just because I escaped any negative fallout from ignoring all the good advice you just received doesn't make it worth it. Let's just pretend for a minute that something did happen. In my case, since Heather and I were the only two members within the LLC, our joint personal assets were on the hook anyway for any litigation, regardless of the LLC. That's right: in hindsight my LLC was pretty much worthless, even if I had treated it legitimately, since we had no employees to be protected from! So instead, say the business had grown to the point where I was able to hire a farm hand, and this employee causes an accident while driving the tractor down the road. The accident was caused by the employee, while on farming business, on equipment owned by the LLC.

You know how this story goes. The victim sees a billboard advertising a less-than-reputable lawyer and goes in for a consultation. She is told that the lawyer will "make 'em pay," and the next thing you know everyone is getting sued—your employee, your business, you personally, the tractor manufacturer for a recalled part that had nothing to do with the accident, the county road crew for not keeping the lines painted properly, everyone. It wasn't your fault, you weren't there, your employee caused the accident, but nonetheless that lawyer is coming after your personal assets as part of his claim.

Will the protections you put in place for just such an occasion stand firm? Are your personal assets truly secure? Have you set up your business to handle this unfortunate set of events and done everything you need to ensure the viability of your livelihood? I know what my answer would have been, and unfortunately it was "no." Here are Ryan's thoughts on how to operate your newly incorporated business in a legitimate manner, ensuring that your answer is "yes":

> Even though the selection of business structures and tax structures for your farm business are tough decisions, the hardest part comes after those decisions are made. That hardest part is operating your new entity in a legitimate manner. Not only is this legitimacy important for liability management, but it is also important for taxes, banking, managing the relationship between co-owners, and proper corporate governance.
>
> The following are some key steps to operating a legitimate business. Some of these steps can be ac-

complished on your own, while others may require the assistance of legal counsel or a tax professional.

Register your business

If you elect to form a partnership, corporation, limited liability company, or a trade name, the business must be registered with the proper state agency. Each state has its own registration requirements, which generally include designation of an agent for service of process, denoting a term of existence, outlining a corporate purpose, and selecting a name not already in use. Be careful to select a name not already in use within your state, otherwise the registration may be rejected. Fees typically apply to any registration.

Draft corporate documents

Aside from using corporate documents as proof of legitimacy in your business, they are essential if you are in business with other people. Corporate documents address areas such as shareholder buyouts, election of directors, powers of managers or partners, dissolution, meetings, and other governance matters. Properly drafted bylaws, regulations, partnership agreements, or operating agreements can tackle any dispute or issue that arises between business owners.

Maintain corporate records

In addition to drafting bylaws, partnership agreements, or operating agreements, maintaining proper corporate records is key. This can include financial statements, shareholder or membership ledgers, minutes from meetings, and corporate resolutions. Meeting minutes and resolutions are particularly important to preserve the key decisions made by owners. If your business is making a major decision, obtain the consent of all partners, shareholders, and members in writing to prevent future contests of that action.

Apply for a tax identification number

Depending on the business structure you select, you may need an employer identification number (EIN). Sole proprietorships do not need an EIN and instead are identified by the social security number of the owner. Partnerships and single member LLCs may want to obtain an EIN, but oftentimes the social security number of the partners or sole member will suffice. Corporations and multi-member LLCs must have a separate EIN. The IRS operates a free online service for any business seeking an EIN if you would like to tackle this task on your own.

Separate your banking, taxes, payroll, and other matters

Regardless of whether you obtained an EIN for your business, you'll need separate bank accounts, payroll, taxes, workers compensation, and other financial components. Overlapping personal bank accounts and business expenditures is one of the top reasons corporate protections break down. Use your business EIN to set up a business bank account and operate all company expenditures through that account. Do not use that account for any personal reasons. Also, remember that some businesses require the filing of separate tax returns. Be sure to follow those filing requirements. Finally, if you have multiple businesses, each one should have a separate payroll.

Update insurance policies

Revisiting Paul's tractor accident example from earlier, your first line of defense is always liability insurance. Forming a corporate entity is a secondary form of protection, with proper insurance being the primary. Upon forming a business, visit with your insurance agent to determine the proper liability insurance amount, the naming of additional insureds, and the applicability of appropriate policy riders such as environmental liability, product liability, executives and officers, and other coverage areas.

Pay rent

This one might sound strange, but the payment of rents is another argument in favor of legitimate business operation. For example, oftentimes a farm business will operate on land owned by the business owners. In this instance, the farm business should pay rent to the landowner for the use of tillable ground, pastureland, farm buildings, or other land-related assets. The same would apply to machinery used by a farm business. Again, this is basically like taking money from your left hand and giving it to your right hand. However, it is an important step towards proving the business operates separately from your personal assets.

Single-member LLC issues

As a final point, single-member LLCs are a unique enterprise when it comes to legitimate operations. Earlier, I mentioned that the IRS treats single member LLCs as disregard entities. This means that the owner and the business are no different from a tax standpoint. Some jurisdictions employ a similar approach for single-member LLCs as it relates to liability protections. This means personal assets may be at risk if an accident occurs involving this type of entity. If you are considering the formation of a single-member LLC, check with your legal counsel to ensure that liability protection will be available to your business.

Get the most from your business structure

Do you need to perform all these tasks in order to properly operate your business? Probably not, but each step you take is one more argument against the plaintiff's attorney from that accident. Some measures, such as drafting corporate documents, separating bank accounts, and updating insurance policies, command more attention. Nonetheless, if you are willing to expend time and resources to incorporate, you have every incentive to maximize the protections and benefits of your business. Take the steps outlined in this section and you will have a strong chance at defending any cause of action that comes against your business.

Talk about being humbled! I can't say enough about surrounding yourself with people who are smarter than you, so that you can go about the honorable business of raising amazing food for your customers while wrapped in the (relative) security that lawyers and tax accountants provide. If there were two areas that I would absolutely recommend resisting your DIY nature, instead paying for the extremely valuable and important guidance that others can provide, it would be here: laws and taxes. It won't make you bulletproof and it will cost you money, but it will also help you sleep at night and keep a single mistake from torpedoing your entire financial future. That's some solid advice right there, take it from me.

Whoops, I mean take it from Ryan...

Write Your Own Story

What goals or needs do you have that would be served by incorporation? Make a list of pros and cons for the different corporate structures—first to decide whether to incorporate at all and then which structure best fits your needs. Jot down contact information for agricultural lawyers in your area, or who to call for recommendations.

CHAPTER 6

Understanding the Marketplace

Thus far, we've talked a lot about all the ways that you can spend your hard-earned money. For the sake of balance, let's spend some time talking about how you'll make money! For the sake of your future business plan, as well as for your personal mental wellbeing, you'll need to consider some very important questions: How will you ultimately sell your farm products? To whom? Through what avenues? This chapter will discuss some of the innovative ways that farmers are reaching consumers, as well as some pricing rules of thumb. Before we really get started, it will do us all some good to step back and define a few terms and ideas. For those of you who have your master's degree in finance, this chapter will be wildly over-simplified, but my goal is to distill realistic options down to a few key ideas. With that in mind, I generally lump marketplace options into two categories: direct sale or wholesale.

Direct Sale Versus Wholesale

My definition of direct sale is any method that involves only the farmer and the consumer. There may be other ancillary players involved — the farmers market collecting a vendor fee for the right to sell at their location is one example. By and large, though, direct selling to a consumer means that there is a straight line between your farm and their table. Wholesale adds a layer between those two—for example, a grocery store buying product from the farm then reselling it at a profit to the consumer. The core delineator for my definitions is money: does it go directly from the consumer to the farmer or is there an intermediary who takes their cut before passing the rest along to the farmer?

I am a huge proponent of direct selling, and count it as one of the key reasons my farm was so successful. I did all the work raising my animals. I assumed the risk of flood and drought. I absorbed the impact of lamb predation. I transported these sentient beings to the processor, then picked up and stored the resulting meat. After all of that, why on earth would I want to share the profits with anyone else? If my farm was to survive and thrive, I felt like I needed to collect as much of the consumer dollar as was available to me.

However, I'll add three caveats to the preeminence of direct selling:

1. First and foremost, it is exponentially more work. By deciding that I wanted to sell at a farmers market, I was also deciding that I needed to spend one hour getting prepared on the farm, one hour driving to the market, one hour setting up, four hours standing and selling, thirty minutes packing up, one hour driving home, and thirty minutes unpacking at the farm... nine hours of effort, every Saturday, just to attend and sell. And when I mean every Saturday, I mean EVERY ONE, no matter if your boy has a baseball game, or your daughter has a birthday party to attend; you best not miss a single market. All. Summer. Long.

2. Exponentially more work, yes... but also exponentially higher costs. Taking the farmers market example (which incidentally is just one of the direct marketing options that we'll discuss): I needed a tent, tables, chairs, portable freezers, a generator, fuel, signage, and display items. In Ohio you have to pay for a mobile meat license, as well as the vendor fees of your market. And you need time. As I mentioned, the time required not only represents effort or work directly applied toward selling, but also money in the sense that we all have a set number of hours each day and how you spend those hours directly relates to your net income.
3. My third caution regarding direct marketing requires some self-inspection. Not the kind when you find a tick crawling on your neck, but one that looks at your own personality and asks the question: am I the right kind of person to stand in front of people and tell my story? Repeatedly? Week in, week out? Am I a "salesman"? The harsh reality is that there are plenty of farmers out there who have no business interacting with the consuming public. I pray that you become an amazing practitioner and a knowledgeable steward of your land, but if you can't handle people, conflict, repeated questioning, incessantly telling your story, and routine rejection, then you either need to avoid direct selling altogether or find someone else who has the right personality to do it for you.

In my mind, wholesale offers an alternative to those who can't, shouldn't, or won't direct sell. By adding in an intermediary, you buffer yourself from many of those pesky consumer interactions. You still have a customer to please and answer to, but they tend to be a lot more informed, direct, and clearheaded. You also don't have to spend all the time and money pursuing the customer directly. However, your brokers also need to get paid, and that money almost always comes out of the farmer's pocket. Sometimes a wholesaler can secure additional margins through superior marketing or additional outlets and pass some of that back to the producer, but more often than not the price for dealing with a wholesaler comes out of the farmer's profits: usually 20-30 percent.

The first question to ask yourself as you consider the options for your farm is "should I sell directly?" Do you have the personality, patience, and demeanor that it requires? Or more importantly: does someone on your team? If the answer is no, then please be honest with yourself and avoid direct selling...you will not be successful, and it will have nothing to do with your skills as a farmer. If the answer is yes, then the next questions is "do I want to sell directly?" Do I have the time to spend on direct sales? Are the extra financial margins worth the extra effort and time?

I am a firm believer that the extra 20-30 percent gross margin that is secured is worth the extra time, effort, and expense. And most of the time I enjoyed the interactions with customers and found value and purpose in helping to connect consumers with their food. However, there are alternatives that might better fit your timeline, bandwidth, and expense sheet, as long as you are willing to accept a smaller piece of the proverbial pie. Also understand that this doesn't have to be an "either/or" decision, and you absolutely can pursue a multitude of options as long as they fit your capabilities and desires. Regardless, consider your options carefully, as the decisions made now will determine where the bulk of your seed money and effort is placed during your startup phase.

Direct Selling Options

Some of the most prevalent direct marketing options include farmers markets, CSAs, on-farm pickup, delivery, and buyers clubs. This is by no means an exhaustive list, so feel free to innovate and think through alternatives that fit your situation! We'll talk about the positives and negatives of each specifically, but in general remember that selling directly to consumer offers as close to 100 percent of the consumer dollar as you can get, but it absolutely costs you more in money, time, and effort. I still believe it is worth it, and have built my successful farm on this model, but you'll need to decide for yourself which model you want to pursue.

A word of advice: whatever option you choose to market your products, make sure it is what you want from the very beginning. The customers you will find at a farmers market are different folks than the ones that will respond to your CSA advertisement, who are different than the ones who will come up your driveway to purchase products. This means the customer base you build in one market isn't necessarily transferable to other markets. Yes, there will be some overlap—those diehard fans who will not let anything stop them from sourcing your product—but my point is that you cannot get started at a farmers market with the thought that you'll get that clientele hooked on your products and then pull them away from the market to a buyers club. They don't come to the market for you; they come to the market for the market. You just happen to be a happy addition to their market world, and they are not likely to leave the market for you. Make sure that the marketing avenue you pursue is selected with open eyes, without any underlying motives to secretly shift people toward how you really want to sell to them. If you really want to sell through buyers clubs, then don't waste your time attending a market. Instead, use that nine hours per week to build up your buyers club model and make that as successful as possible.

Now let's explore some of the direct sale opportunities:

1. Farmers Markets

The first thing that probably comes to most people's minds when you say "direct sell" is farmers markets. They are the quintessential common grounds where producers and consumers gather to trade food for cash. Seldom is there a situation where you can quickly access an established group of customers for such a small price, yet sell to them directly. No commission, no consignment, just a "pay to play" approach to getting you in front of a group of buyers. Among the perks of selling at farmers markets are that markets will often shoulder much of the burden of marketing and are constantly looking for innovative ways to showcase their vendors. Committed buyers will begin to seek you out weekly, plus you gain a ready-made location for non-market customers to pick up pre-sold orders.

My biggest critique of farmers markets are the relatively high expenses required, including but not limited to: vendor fees, mobile licensing, liability insurance riders, signage, generator, tables, chairs, display items, and the highest expense of all: time. Some of these expenses may vary based on your location and set-up, but all are worth considering as you make your decision of whether to pursue this market.

In addition to these very tangible marketing and advertising expenses, taking the step into a farmers market also represents a large mental draw. Consumer education becomes your main task—explaining your production methods incessantly, constantly walking the line between advocate and salesman. Consumers' base knowledge regarding food production varies widely, as do personalities. Some are there to shop, some to see and be seen, others troll for last-minute deals. When you attend a farmers market, you open yourself up to all kinds of food shoppers; you need to be prepared mentally to interact with all of them.

I remember a crotchety old woman who approached my booth one morning and asked if I still had eggs available. True pastured, non-GMO eggs are essentially pure gold at any farmers market, so I was surprised at the time to be able to say "yes"! I went to the refrigerator in my trailer to grab one of my last cartons and, as was my custom, opened it in front of her to show that none were cracked or damaged. She reached into the carton and pulled out an egg, eyeballing and inspecting it, holding it up to the sun as if it was actually a gold nugget. Then she put that egg back into the carton…and picked up another one! This turned into one of the most awkward and lengthy market interactions I've ever had the misfortune to be a part of, as she picked up and inspected every single egg in that carton. Finally, literally after touching every egg, she looked up at me and sneered "These look like pullet eggs" before turning and walking away!

Still want to sell directly to the consumer? To be fair, for every negative story I have there is at least a positive one to balance it out. A few years back, I was shyly approached by a young lady who asked "does your meat have any additives or fillers in it." My enthusiastic answer was "absolutely not," so she tentatively bought a package of ground beef. The next week, the same woman came back to me, but this time she was glowing, bubbly, and excited, exclaiming "I can eat your products!" As it turned out, this woman had an autoimmune disorder that had prevented her from eating normal proteins and meats, likely because of the genetically modified ingredients in the feed of the grain-fed animals. She hadn't had meat in over a year, due solely to the extreme negative reaction her body had to the grocery store options, reactions that did not occur when she found my clean, healthy, unadulterated products. She became a loyal and dedicated customer and was even able to eat other products of mine like eggs, pork, and chicken, which, according to her doctor, shouldn't have even been an option for her with her condition. If I hadn't made myself available to her, and other customers like her at the farmers market, it is doubtful that she would have ever found me and experienced the freedom that good food offers.

One customer experience like her erased all of the egg-inspecting octogenarians that I had to deal with, making the farmers market an enjoyable and viable option for me.

2. Community Supported Agriculture (CSA)

The Community Supported Agriculture (CSA) model has gained massive popularity in recent years, although it also has morphed into something altogether different than its original intent. The concept is that customers pay their farmer up front at the beginning of the season, solving part of the eternal agricultural issue of seasonal income. In return, they share in the bounty of the season in good years and take the hit alongside the farmer in the bad years. Often there is some sort of minimum expectation for what the customer would receive over a season, with the hope that abundance would flow well above that minimum threshold. A weekly pickup, either on-farm or at a common pickup location, is typical.

While admirable in theory, too often the CSA name is taken in vain as customers demand that real farmers conform to faulty expectations based in grocery store aisles: convenience, variety, perfection, and regularity. Bending to that will, farmers stop asking for money up front, start allowing subscribers to pick and choose, and use pre-orders to create custom boxes. All of that is fine, I suppose, if that is the model you wanted in the first place, but it isn't a CSA...the protection for the farmer, the shared risk and understanding by the consumer, and the connection with the realities of farming are all gone.

CSAs generally lend themselves to produce and fruit farmers and tend not to be as applicable to meat ranchers such as myself. However, there are still unique opportunities to partner with an existing CSA, taking advantage of their established customer connections while offering an add-on product that would help them differentiate themselves from their competition. I see this type of relationship as holding the most potential for success, at least for us meat and egg folks.

3. Buyers Club

What the CSA model has mostly changed itself into is a buyers club, a model that I see a large potential for success with. My issue with so-called CSAs isn't what they've become, just that they are no longer what they say they are. A buyers club model involves the farmer collecting pre-orders from a geographically targeted set of customers, then meeting and delivering those orders at a common location convenient to the customers. For this, you can use a customer's home (and offer them a discount for the use of their address) or, in the irony of ironies, a grocery store parking lot—or really anywhere in between. For my buyers clubs I used an outdoor mall's parking lot, meeting customers in front of the closed credit union.

There are many positives with this model. I spent one hour getting prepared on the farm, one hour driving to the meet-up location, allowed for a gracious two-hour window for pickups, then drove one hour back to the farm. And since I only offered this twice a month, my time commitment was just five hours every other week compared to the nine hours per week for the farmers market. No set up, no tear down, and if I didn't have any orders for a scheduled day then I didn't go! If all of my customers happened to arrive in the first 30 minutes of the two-hour window then I drove home early. I still had the benefit of connecting directly with my customers, sharing my story and what was going on with the farm. And I only carried product that was already sold—no need to keep inventory cold for hours at a time as if I was a supermarket.

Obviously, I'm a big fan of the buyers club idea, but it does have its pitfalls. One is marketing, with the obvious question being "how will they find me?" Often that is why farmers start going to farmers markets, in order to get our name out there. But then we get trapped in that world; we've spent hundreds of dollars on the marketing set up, so we might as well stay there. Instead, spend that money on a website presence, develop your email list, and leverage both your social media and your actual relationships (more on all of those in Chapters 8 and 9) to gain visibility. Another downside is the specter of driving an hour one way because a single customer ordered a dozen eggs. I tackled this by requiring a minimum $100 order for the trip, in order to ensure that it was financially worth it. I didn't care if that minimum order amount came from one customer or ten, I just knew that at least my gas and four hours of time were covered if I was making the trip. Incidentally, I did offer free delivery with no minimum order, specifically restricted to my local zip code. I found that I could usually fit these deliveries in along with errands or kids' appointments without any problems, and they were worth the small cost to develop and maintain local relationships with customers.

4. On-farm Pickup

This leads me to the final direct-marketing option that I will cover here: on-farm pickup. There is power in allowing customers to come to the farm, but also liability. Part of the reason direct marketing works in agriculture is because our customers, whether they recognize it or not, are pursuing a connection with their food. They are opting out of "normal," instead driving up my dusty driveway, looking out over lush pastures and happy cows, envisioning the bucolic lifestyle that they so desperately want to support.

Offering that to your customers is the least you can do, but be forewarned: they aren't going to want to leave! Each visit turns into a longer-than-it-needs-to-be visit to pick up a pound of ground beef—a pleasant chat, to be fair, but one that gets in the way of the thousand things that need done in that moment. Heaven forbid they arrive at the moment the water line breaks, the cows get loose, or the storm is coming. Bottom line, let people come to the farm, but do it on your time. Have "office hours" where you know the potential of interruption exists. And if you need to be unavailable, then be unavailable. It's your business and it doesn't do you or your customer any good if you can't run it appropriately. Set boundaries and stick with them, and don't be afraid to say "not today."

Selling directly to consumers offers the ability to maximize your gross margins by securing the highest percentage of the shopper's dollar, but it is by no means cheap or easy to accomplish. You have to have the right personality to handle that face-to-face interaction, and you have to pursue the right marketing model for your situation. Don't fall into the trap of marketing one way, just because it is convenient, if it doesn't match your vision for your farm or fit into your fiscal, personal, and business constraints. Instead, decide how you want to sell and spend your limited resources making that model successful.

Wholesale Options

"Wholesale" is defined as "in large quantities, or extensive in nature," generally indicating a larger-volume sale. Keep in mind that that doesn't necessarily mean more products sold overall and almost certainly doesn't mean more income; it just means a larger quantity of goods per sale. The gist of wholesale is that you as the farmer sell your products to someone else, who then does something to them: cook and prepare, value add, distribute, deliver, or stock their shelves, and then they sell your products. Those are all good things, but as I mentioned earlier, they all cost money, and more often than not that money is partly "eaten" by the farmer. It may still be worth it for you to consider, but there is a fine balance that needs to be struck between sales volume and customer reach versus gross income and production costs.

For the purposes of this conversation, I am going to lump restaurants and grocery stores together. The positives and challenges that I discuss here are the same for both, but recognize that there are at least as many differences between the two marketing outlets. I have much more experience with restaurants, which explains the emphasis that you'll see, and I will attempt to highlight applicable differences between the two as they arise. Additionally, the pricing examples I'll use are real and are from my farm, but recognize that they are highly localized and should not be used to base your product pricing on. After we discuss wholesale options, I'll close this chapter by talking about pricing your products for market. Bottom line: don't use my numbers as a baseline for your projections.

Here are some of the wholesale opportunities available for farmers:

1. Restaurants / Grocery Stores

I have sold through several wholesale channels, and my most pleasurable experience was selling to a local restaurant in Chillicothe called The Green Tree. The owners saw value in highlighting a local connection to the food they served and were already buying grass-fed beef from their weekly delivery company. We struck a deal for me to become their exclusive ground beef provider, after which they used my product for burgers, shepherds pie, chili, and other seasonal specials. I have to say, it is really exciting to sit down at a local restaurant and see your name and logo on the menu. There is a feeling of hitting the big time, like somehow you've really made it. The same would be true, I can imagine, of scanning the meat aisle at the local grocery store and finding your branding proudly and prominently displayed.

When the owners of The Green Tree and I sat down to hash out our deal, I ended up offering my ground beef to them for $5.25/pound—almost 50 percent off my retail price of $10/pound. Why would I do that? There are several reasons. Of course there is the value of that local connection, my logo and story in front of Chillicothe diners every night, as well as cross-promotional opportunities. That said, never did I have someone come buy a side of beef from me because they had a burger at The Green Tree, loved it so much, and got my contact information from the server — not once. The main reason I lowered my price that much was because I was using the restaurant to move cull cows that weren't much good for anything else besides ground beef; they were too old and too tough. My only other option would have been to sell them at the local sale barn. The sale barn price would have averaged $750 per cow, and was also highly variable and unstable thanks to the commodity system that most farmers sell through. Instead, by selling to The Green Tree, my net income per cow was around $1,450 and was stable and predictable. Not to mention my meat was going into real food, instead of into the fast food, dog food, and glue channels!

There is a reason that The Green Tree didn't have my filet mignon, ribeye, or chuck roast on their menu, highlighting the first of several issues surrounding selling to a restaurant or grocery store: volume. Did you know that there are only eight to twelve filets and ten to fifteen ribeyes on a single cow? Assume that The Green Tree sold just five steak dinners per night and were open six nights a week. In the best-case scenario, I would need to provide them a cow per week, or fifty-two cows per year, just for that outlet alone! And what was I supposed to do with the rest of the animal? That raises a second issue: specialization.

It takes a very special kind of chef to actively pursue the level of seasonality and diversity that real agriculture entails, and unfortunately those people are few and far between. Instead, most restaurants take a "specialty" approach to their menu creation, serve a specific niche of the dining public, and keep their menu full of old standbys and crowd favorites. There is one such restaurant here in Chillicothe called the Old Canal Smokehouse. They are renowned for their smoked meats and delicious barbecue, and they even have a really good smoked burger! A few years ago, I met friends there for lunch and said to myself, "Hey, I raise pork—why not try to serve this restaurant?" I asked to meet the owner, with the intent of gauging his interest in working with a local producer. As it turns out, he was really excited about the idea, and immediately started talking about all the positive things we could do, how we could market, and how his customers had been asking about local sourcing! At this point I was really getting hyped up about the endless possibilities and potential this new relationship represented, up until he uttered the words that took all the wind out of my sails: "I need six pork shoulders per week."

Wait, what? Just the shoulder? What about the rest of the animal?

"No, I don't smoke anything but the shoulder, so that is all I would need from you."

Needless to say, I had to awkwardly end the conversation, leaving the restaurant dejected and dismayed. How could a small farmer like me compete against the convenience of buying six pork shoulders every week from some massive, faceless, CAFO hog barn that delivers right to their doorstep? Restaurants (and even some grocery stores) often have this level of specialization, because they are told they need it to maintain their customer base, the cheap food system that supplies them allows it, and the American eater psyche demands it. It is an unfortunate reality that the concept of an entire animal dying and the preeminent need to utilize the whole animal is a foreign concept to most restaurants.

Even if I could have somehow raised and processed that many hogs for Old Canal, or that many cattle for The Green Tree, what would happen if they sold really well and needed more product to meet demand? That question raises the third issue: volatility. Restaurants are on the extreme side of dealing with the whims of the consuming public. Yes, they can track sales data to predict an increase in patrons around major summer holidays. Yes, they can shift their menu, offer specials, and create demand through pricing to help control consumer menu selection. However, their ability to hold and store a large stockpile of products is severely limited, especially considering that they have to have all the raw materials in stock to theoretically make their entire menu every night. It's no wonder that ordering small quantities off a routine delivery frozen food truck is their norm. Even with just ground beef, I often found myself storing meat on my farm for The Green Tree because they didn't have the freezer capacity. As much as possible, I helped them plan ahead, but on one occasion we misjudged and I didn't have any beef animals available to process. Their supply ran out and, unbeknownst to me, they backfilled with beef off the delivery truck for several weeks until I could resume delivery. The problem was that they never said anything, either to me or to their customers. I was livid, but it's hard to hold it against them. What were they supposed to do—completely reprint their signage or take burgers off the menu for several weeks?

Obviously, dealing with restaurants or grocery stores could offer you the ability to move lower-end bulk products or provide a safety outlet for excess products, but dealing with chefs and food managers is an entirely different marketing arena than a farmers market. If you choose to pursue a local restaurant or grocery, here are some tips for success:

1. Plan your visit to avoid their busy times, being respectful of their schedule
2. Understand their menu and what you plan to offer them
3. Take samples to share
4. Be prepared to speak to your products' freshness and quality —including some education on how it was produced/raised.
5. Don't minimize the value you are presenting—you and your product could be an important part of their brand!

Generally, restaurants are running on the ragged edge of profitability, averaging just a 3-5 percent profit margin. Because their margins are so tight, you can expect to sell them products anywhere from 20-30 percent off your retail price to account for their overhead costs. Whether or not that cost of doing business will pay for itself in volume, cross-marketing, or additional connections is a tough question, but restaurants seldom have much wiggle room when it comes to what they can pay you for your hard-earned product. For the purposes of contrast, let me present a different way to increase volume and product offerings while addressing the issues of volume, specialization, and volatility: perhaps joining a cooperative would be worth considering.

2. Farm Cooperatives

A cooperative business model is "a private business that is owned and controlled by the people who use its products, supplies or services." In agriculture, feed stores, grain mills, and supply outlets are more often legally formed and operated as cooperatives. But what I am referencing here is the concept of a group of farmers or ranchers producing under a shared standard (and maybe a shared brand) for the purposes of accessing larger value and volume markets. I see a great deal of promise for this version of wholesale marketing, in that it pools resources, spreads risk, and accesses unique personal skill sets that might otherwise not be available to a single-family farm.

Let's say Farmer Paul, Farmer Ryan, and Farmer Sarah got together and decided on a shared set of production standards. For example: 100 percent grass-fed and finished, no routine antibiotic use, no growth hormones, exclusive use of OMRI-approved products for livestock and land improvement, humane handling procedures, and minimal transport times to slaughter. In addition, Sarah happens to be an excellent marketer, while Ryan's skills lend themselves more towards accounting and bookkeeping. Neither Ryan nor Sarah have any hay equipment, but Paul does, along with stock and equipment trailers. Instead of each one of them spending the time and money to create their own branding, invest in specialized equipment, secure a land base, and eventually compete with each other, they pool their resources within a shared brand, take full advantage of each others' strengths, and bring a large volume of product that can supply a restaurant or grocery store market year round.

That sounds like a recipe for success! Multiply this concept from three farmers and make it thirty, and perhaps a cooperative group could secure a mobile processing unit and collaborative cold storage, addressing supply-chain limitations and satisfying consumer demand for humanely slaughtered meats. I believe there is a massive potential within the cooperative model, but it does fly in the face of the individual, self-reliant, stalwart image that we farmers have for ourselves. In addition, it requires an extra layer of legality in the sense that issues like income distribution, shared responsibilities and tasks, liability, and exit strategies need to be clearly defined and agreed upon. Arguably, we should all have those things figured out anyway, but joining forces with others in production and marketing would mandate that level of detail and structure.

If there is an existing cooperative in your area, check them out online or make a phone call to discuss their needs. You might find that, much like my "vegetable CSA needing a meat add-on" example, existing cooperatives may have a niche food need that is currently unmet from within their membership. Sometimes cooperatives aren't farmer-oriented at all, instead freely buying and selling from independent producers around them, and would be happy to add your product to their local food lineup—at a wholesale price, of course.

Whether you choose to pursue a restaurant, grocery store, or existing cooperative as an individual enterprise or as a cooperative yourself, recognize the fact that you are adding other entities into the supply chain that stretches between you and your customer. That might offer increased opportunity and/or volume, mitigate personal shortcomings that inhibit your direct-marketing opportunities, or address physical limitations like time, availability, and distance to market. But it will certainly cost you on the income side, as you absorb some of the overhead costs of those additional links in the chain. Speaking of income, now seems like the right time to broach how to price your products in the first place.

Pricing Your Product for Your Market

Putting a price on your farm product is a highly emotional experience for us farmers. The price you ask encapsulates so much more than profit; it includes all the sleepless nights you spent listening to the coyotes howl and wondering if your lambs were safe, that late frost that caught you and your tomatoes off guard, the equipment breakdowns in the field, and the love, care, and stewardship you poured into your operation to bring that specific tomato or that pound of ground lamb to your customer. The problem is that your customer doesn't understand that, and despite how much we tell them, they never truly will. Therefore most conversations you hear about product pricing revolve, as this one will, around mechanical, mathematical, logical approaches. However, it is worth recognizing from the very beginning that price represents so much more than the result of a formula. What you need to figure out is how to capture your sleepless nights and stewardship efforts in order to arrive at a price that you can be proud of and your customer can afford.

My first and strongest piece of advice speaks directly to the affordability question: DO NOT attempt to be a low-cost producer; that is a fool's game. There are generally two categories in any type of business: low-cost or high-value. Whether you are purchase clothing, tools, or wine, all of us have a decision to make. Do I spend money on a quality product that will last longer / taste better or do I cheap out and buy something that will disintegrate in the washer, break in the field, or taste like turpentine? I would argue that small-scale farmers fit squarely in the "craftsman" category of goods and services, and our pricing needs to reflect that. Trying to produce food cheaply is a downward spiral that looks remarkably like a toilet flushing, and that approach is what has gotten America into the sewer of our current food system. Many people point to innovation, technology, efficiency, and scale as ways to reduce or spread cost across more products, therefore reducing the cost of production, and I am not against any of those things. You can produce high-quality food on a large scale—just look at Will Harris at White Oak Pastures as one example. You can and should utilize technology to be more efficient. But there is only one way to produce cheap food: you cheat. Sub-therapeutic antibiotics, growth hormones, genetically engineered seeds, CAFO animal factories—these are all methods that indicate a business has chosen profit over values in an effort to produce food as cheaply as possible. Don't be one of those farmers; your customers, neighbors, and family deserve a high-quality product worthy of paying good money for.

So how do we price items? How much is it worth? How much is too much, because we have to sell it to make money, right? It is so tempting in the face of those questions to glance across the aisle at our competitor's price board, look up someone else's online shopping platform, or find similar products in the grocery store aisle and just copy their prices. Surely they've done their homework, analyzed their markets, and logically arrived at that price. Perhaps. Or maybe they did the same thing you are now contemplating, and so did the farmer that they copied from, and the chain of uninformed pricing goes back multiple iterations!

That's a cycle you don't need to perpetuate. But while I don't want you to copy your competitor's prices outright, it is worth knowing what they are for two reasons: it serves as a reality check on what you come up with independently and it provides you with the knowledge that you are either cheaper or more expensive than they are in order to inform your conversations with your customers.

The good news is that we don't have to copy—or, even worse, guess—to price our products. Instead, you need to start with what it cost you to produce said product. Ensuring that you are priced at least at/above cost of production seems so obvious, but you'd be surprised how many farmers don't do it. It goes back to one of the initial questions I asked you: are you running a business or is this a hobby? Nothing wrong with either, but a business needs to be profitable to survive; a hobby does not.

I am a strong advocate for something called "enterprise accounting," which is a process where you break down your accounting in your overall operation by the individual enterprises. We will go over this in detail in Chapter 10. For the time being, let's just cover the basic costs of production with a simple pork enterprise example, where I am purchasing ten weaned piglets, raising them to finish, and processing and selling the meat. To bring that pork to the point of sale costs me:

$500	Purchase 10 piglets, $50 each
$2,070	Feed, 9000 lbs
$2,000	Processor, approximately $200 each
$4,570	**Total cost**
$457	**Total cost per pig**

So all I need to do is make sure that my cumulative pricing for one pig's worth of products covers the $457 cost of production... what's the big deal? No wonder they call hogs the "mortgage lifter." Hogs are so profitable!

Not so fast. Let's take a step back and look for what is missing from this example. For starters, we're definitely missing some of the more hidden costs:

- Infrastructure (fencing, feeders, waterers, barn, trailer, freezers)
- Utilities (water, electric)
- Transportation (fuel, mechanical, for both initial pickup and processor trips)
- Marketing (online presence, farmers market fees, mileage)
- Death loss (minimal for hogs, but not for a poultry example!)

As you can see, there is much more that goes into producing a pound of bacon than a simplified chart. However, even after we are able to track down each individual input that needs to be accounted for, I'm willing to bet that there is one more thing that we've forgotten.

Farmers are some of the worst about this; we always forget to account for...ourselves. We almost never pay ourselves, but our time is valuable and must be accounted for. Daily chores, time spent in the truck moving animals or picking up feed or getting to the farmers market, time spent in front of a computer building a pork products flyer. However you tally everything up, your time had better be part of your calculations. Personally, I value my time at $30/hour.

Another way to handle this would be to add in a "profit margin" above production; if you take this route I would recommend 25-35 percent for your high-end, craftsman-level product. Adding in those hidden expenses and our effort changes everything. Without sharing the entire cost chart and how I came up with it (again, an entire discussion on enterprise accounting awaits you in Chapter 10), my true cost of production for a single hog was $840—almost double the simplified initial estimate!

Now that you know the costs of production, the next question becomes how much meat comes off an individual animal? As I mentioned earlier, once I started tracking and adding up the cuts coming back from the processor, I was shocked at how few premium steaks there actually are on a finished steer. To figure this out, you have to track and count up the weights you get back in order to come up with an average number of each item. For my hog example, I figured on 32 pounds of sausage, 46 pounds of loin chops (bone-in or boneless? It matters), 12 pounds of bacon, 6 pounds of spare ribs, 24 pounds of ham, and 10 pounds of shoulder, which totaled 130 pounds of meat. That means that my average price per pound for pork should be at least $6.47 to cover my true cost of producing that hog. There is definitely flexibility to move that price around depending on the cut, and this is where you can start to compare yourself against your competitor. She sells her bacon at $8 per pound? Maybe you can match that, but only if you sell your sausage at a pricey $6.50 per pound. Which will you have more of? Which is more likely to be seen as a premium cut? Probably move the sausage at $5 per pound, sell the bacon at $12, and be ready to tell your customers why the price difference is so completely worth it!

That brings me to my final word of advice on this topic: don't back down. You will have worked hard on your product. You put in the time, you put in the money, your animal or plant paid the ultimate sacrifice, all so that your customer could eat healthy, clean, and humanely. That is worth the price, no matter how many times someone visibly blanches or outright scoffs at your prices and walks away with their nose in the air before getting in their car to wait in the fast-food drive-thru line. Be ready to tell folks why your product costs what it does, but do not waiver, regardless of whether you decide to pursue wholesale or direct-sale outlets. If you can't sell your product for what you need to, then the only conversation that needs to happen is whether or not to continue that enterprise. Have a sale every now and then, offer excess inventory for a reduced price or through a wholesale market, adjust or shift your prices based on demand and supply, but always keep your bottom line in mind and do not sell below it—that is the realm of a low-cost producer who cheats to sell cheap food, and that cannot be you.

Write Your Own Story

Describe your ideal customer...where and how do they purchase your products?

List your available time on the left side of the page, then your potential market outlets across from them on the right side. Draw lines connecting the market outlets that match well with your current availability. Is there a market that you need/want to meet that might require an adjustment of your time?

CHAPTER 7

The Web of Regulation... Is It Really about Food Safety?

The government is here for our protection. I promise.

Except when it comes to food. Then it often appears that the sole purpose of our multi-level government bureaucracy is to harass and harangue farmers, making our job as difficult as possible. Overreaching, overlapping, contradicting, one-size-fits-none regulations are something you will just have to deal with in your new role as "producer." I hope you have thick skin and a bulldog-like personality, because you're going to need it! While there are definitely some bad apples and power-tripping naysayers, most regulators are actually very nice people with good intentions. The problem is that most of the time you don't get to choose which kind you have to deal with, and often the regulators are just as confused on the rules as you are.

Not exactly a vote of confidence, you say? That's an understatement, and I'll get to my reasons why that is, but I wanted to start this chapter with the recognition that most of the folks that you'll come into contact with in the regulatory realm are overworked, underpaid, and frustrated with their lack of ability to help. So give them a break, even as you push hard for some common sense within the vast and confusing web of regulations that lies ahead of you. And push hard you will have to, because no one is going to advocate for your point of view except you. You'll need to find creative solutions to the rules that you can't avoid completely, obeying the letter of the law while bending the execution to your intent.

Overreaching Regulations

In case you couldn't tell where I stand on food regulations in general: they are mostly burdensome and unnecessary. Don't take that to mean that I advocate for removal of all food safety regulations—absolutely not! We need rules in place to ensure the safety and security, distribution, and sale of food products. Nothing I say here should be construed to mean that regulation has no place in our food system. While I have to believe that all regulations must have, by necessity, addressed a real problem at some point in their history, many have lost their way and become part of the problem instead of part of anything that resembles a solution. It's almost as if sometimes the rule and the need for it don't connect; often they aren't even close!

Most people wouldn't argue with me if, for example, I said that food needed to be kept cold to prevent spoilage. Except for eggs, but we'll talk about that little gem in a bit. Generally speaking, colder temperatures keep bacteria that is naturally on our food from multiplying and taking over, as well as slowing the decay of living cells. That is why properly packaged frozen meat can keep for years, and lettuce lasts for a couple of weeks in a crisper instead of a couple of hours lying on your kitchen counter. There are thresholds established for frozen and refrigerated foods to be legally sold, which also makes sense to me. You can't enforce a rule that doesn't have a number associated with it, and theoretically that number was arrived at after much testing on the decay rates of common foods instead of being picked arbitrarily by a bureaucrat. Yeah right, but regardless of how someone arrived at the numbers, 41 degrees Fahrenheit or below is officially "refrigerated" and 34 degrees Fahrenheit or below is officially "frozen." Food kept within those temperature ranges are safe to transport and sell for human consumption. Food safety is important—got it.

The rub comes when a regulation not only requires that we keep food cold, but HOW we keep food cold. In my case, this regulation is at the county level, and Ross County, Ohio, requires its food producers to utilize commercial refrigeration units. These refrigerators and freezers come with a special sticker denoting their commercial status, right next to the price sticker that is easily double the non-commercial unit sitting next to it. Theoretically, commercial units are more robust than non-commercial ones, but that's not really the point. The point is that the regulation just crossed the line and no longer pertains to food safety. If I can keep my products within the safe temperature ranges, then what concern is it of anyone's how I do that? Ice and a Yeti, a 30-year-old chest freezer, or a shiny stainless steel commercial freezer—it shouldn't matter; 34 degrees is 34 degrees. Safe is safe.

I had already purchased my freezers to install in my market trailer before I found all this out. I got a great deal on several used and slightly rusty units on Craigslist, but who would care because they worked, right? Wrong. After calling the county health department to schedule my first inspection, I was informed that all of my freezers were illegal because they didn't have the commercial sticker on the back. For a brief moment, I actually considered printing my own stickers at home and putting them on the back of the freezers, but cooler heads (hah!) prevailed and I decided against that. Ultimately, the inspector came out and did the inspection, giving me my "mobile food" license despite the non-commercial units. My favorite part of that interaction was when we were discussing my lack of compliance from the very beginning and the inspector told me "Don't worry about it—worst-case scenario is that someone does a spot inspection and writes you up for this. That's not likely to happen, but even if it does it would take multiple write-ups for us to actually come out and do something about it. So you're pretty safe." Great, thanks. Guess what? Those non-commercial units kept my frozen meat frozen for the next seven years and I never got written up for using them. Perhaps I wasn't the only one who could see that the commercial requirement had nothing to do with food safety anymore!

I actually felt sorry for my county health department, because they honestly didn't know what to do with me at the beginning. This hippie, tree-hugging regenerative farmer comes into the picture and wants to do all sorts of things that don't fit nicely into their food-safety box! I wanted to put my freezers inside an enclosed trailer, so that I could park and plug them in at the farm, but easily transport everything to a farmers market without loading and unloading separate freezers in the bed of my truck. The best answer they could come up with was to issue me a "mobile food" license, as if I was a food truck. To be fair, my trailer did have wheels and could move, so maybe that makes sense. We had a little bit of a standoff initially, as part of the inspection checklist for a mobile license involves four-bay stainless steel sinks, potable and hot water spigots, and hand-cleaning stations, which only made sense if I were an actual food truck preparing hot food for sale. I wasn't, but the regulation doesn't allow for variation. It was touch-and-go there for a bit, but in the end we agreed to waive those requirements as long as I promised not to cook anything.

A whole new issue raised its head when my inventory outgrew the available space in the trailer freezers and I needed to store meat in freezers that were not located on the trailer. It was the same physical location, same power source, same environmental conditions, and same non-compliant, non-commercial units, yet somehow if frozen meat moved from being stored on the trailer to being stored in the garage the trailer was parked next to, then suddenly I needed to apply (and pay for) a meat warehousing license! Are you kidding me? Again, frozen meat is frozen meat, safe is safe. This made no sense, clearly had nothing to do with food safety, and represented another overreach of bureaucracy into my world. Therefore, I may or may not have illicitly yet safely stored frozen meat out at my place for years!

You'd think I was a modern-day Robin Hood resisting the established government, the way I had to navigate and sometimes sidestep overreaching regulations. I justified my actions with the knowledge that food safety was never in danger. You might not be comfortable with that, and that would be completely fine. You'll just pay a lot more money and deal with a lot more frustration than I did. The point of this discussion is that you will inevitably encounter food regulations that have lost their way and overstepped their bounds.

Overlapping Regulations

In addition to overreaching, many regulations come with a confusing and overlapping set of boundaries, responsibilities, and gatekeepers. The details will be different in each state, but I guarantee there will be overlap. In Ohio, the county health department is responsible for regulating meat storage and sales, which is why I had to get my mobile food license inspected through them. When it came to storing my eggs however, that fell under the Ohio Department of Agriculture, which incidentally does not require commercial refrigeration—score! That one-time inspection centered on a refrigeration unit that maintained a safe temperature and the source of the water that would be used to wash the eggs. In my case, I was on a municipal water source so I was an easy approval, but heaven forbid someone try to wash eggs in water sourced from a well; then I would have had to provide, at my own expense, water tests that proved bacteria levels below a certain threshold. I'm sure that threshold was scientifically sourced and vetted too, but thank goodness I didn't have to find out!

I promised to come back to eggs, and this story begs the obvious question: why do we wash eggs in America anyway and does that have anything to do with food safety? Most of the developed world actually prohibits eggs from being washed and they are sold unrefrigerated, so what gives? It turns out that when a hen lays an egg, it comes out with an invisible impervious protective layer called the "bloom." The bloom keeps dirt, manure, and bacteria that might collect on the outside of the egg from passing through to the inside and harming a developing chick. In a production system without roosters, this means that all eggs naturally arrive with a protective layer that keeps the food within safe for consumption. How cool is that? As long as you don't wash the bloom off the egg, then you can safely store and sell them unrefrigerated. Other countries recognize this, and the norm is to find eggs for sale on a pallet in the middle of a grocery store aisle.

In America, we chose a different route. In our infinite wisdom and hubris, our regulatory structure mandates the washing of the egg, which in turn makes refrigeration necessary, which in turn benefits everyone except the farmer and consumer: electrical companies, refrigerator salesmen, and bureaucrats alike. Like a self-licking ice cream cone, we now have an example of an overreaching and overlapping regulation that actually creates the problem that it subsequently seeks to solve. You can't make this stuff up!

Contradicting Regulations

Not only do food regulations tend to be overreaching and overlapping, but they sometimes go so far as to appear contradictory. Sometimes that is due to the interpretive nature of regulation, in which far too much power lays with the bureaucrat currently interpreting the rule with their own biases and baggage clearly on display. Other times it is because of the patchwork nature of food regulations, in which counties, states, and the nation all have their own ideas on how farmers should be squeezed...I mean...supported.

While my county health department required commercial mechanical refrigeration for the "safe" storage of meat, other counties in the state had completely different rules. This contradiction was no more evident than at the large farmers markets that I attended and sold through in the Columbus metro area. With more than eighty vendors total, there were inevitably eight to ten meat vendors from all over the state competing for the (relatively) wealthy customer base who purchased their food there. I can remember being confused at first, then progressively more upset, as I took a break from selling to walk through the market and check out the competition, only to see a menagerie of commercial freezers, non-commercial freezers, and coolers filled with ice. Some had generators running to keep their units cold, others just used chest freezers as makeshift storage units with the plugs dangling off the back of their pickup trucks. There was no consistency, nor commonality, among the rules that each county health department required for their food safety regulations.

The worst job in this whole mess had to be that of the local health department inspector, who periodically visited my booth to check my thermometers and make sure I wasn't peddling unsafe products. Regardless of their own set of rules, they were required to know and honor the food safety regulations set by each of our home counties. Can you imagine? It raises the question then: what if farmers got wise enough to shop around for counties that had more advantageous and less onerous regulatory environments? What if you happened to farm in more than one county? Could you pick and choose which office to call for your inspection and licensing? It's an interesting concept, born out of the contradictory nature of food safety regulations, but alas my farm was squarely seated in the middle of my county, so I was stuck complying with their interpretation of red-tape requirements.

This contradiction exists on a state level as well, as evidenced by how state departments of agriculture treat "weird" meat animals. Take rabbits as an example; it's pretty clear that no one really knows how to handle them. I would have loved to raise and sell a pastured non-GMO rabbit. They are a lean, healthy meat source and I believe it would have been a huge hit with my customer base. Processing the animals is a breeze, requiring a simple tool to quickly and humanely dispatch the bunnies, then some quick knife work before peeling the pelt off and eviscerating the chest cavity. A rinse in cold water to clean up the carcass, packaged in either a shrink or vacuum-seal bag, and straight to a refrigerator for cooling. In five-minutes or less I would be able to turn a living animal into a delicious and healthy meat option for my customers. Simply. Cleanly. Safely.

Contrast that process with poultry butchering, during which you kill the animal with two quick swipes across the arteries in their neck, wait for them to bleed out, then dunk the carcasses repeatedly in hot water to force the skin to open and release the feathers. This scalding water is used for multiple birds, before blood and manure make it filthy enough to require changing and reheating. Meanwhile the properly scalded birds go into a tumbler one by one, which bounces them around inside against rubber fingers that remove the loosened feathers. Again, the tumbler is used repeatedly for all the birds being processed, with a quick spray down of water occasionally whenever the feathers begin to build up. Then the now naked birds move to a table for feet and neck removal, evisceration, packaging, and cooling.

I share these two processes in order to provide context for the contradiction they represent. In Ohio, anyone can process up to 1,000 birds per year, on their farm, with no inspection requirement whatsoever. Yet for some reason, the Ohio Department of Agriculture labels rabbits a "non-amenable species," alongside bison, camels, ostrich, and captive deer. As such, they not only have to be processed at a slaughterhouse under state or federal inspection, but they must also undergo a secondary "voluntary inspection" that is in fact mandatory and also costs extra money. This despite the fact that processing a rabbit is a much cleaner process with far less risk of introducing pathogens to the final product. Again we see the obvious question that needs to be raised: is this really about food safety at all?

Compounding my plans for raising rabbits was the fact that very few slaughterhouses have rabbits listed on their Hazard Analysis and Critical Control Point (HACCP) plan and very few inspectors would have any idea on what they were supposed to be looking for in the first place. Needless to say, I was never able to crack the bureaucratic code to make it happen, no matter how many times I called the Ohio Department of Agriculture hoping to speak to someone with a different interpretation of the regulations. Just as it would be an interesting experiment to shop around for county regulatory environments, the same could be true at the state level. Some call rabbits non-amenable. Others lump them in with poultry. Others actually address them in their own right. It's all just part of the contradictory nature of food regulation in America!

Food Safety Modernization Act

I believe in "small government" and a decentralized approach to governance. Seldom does a rule or law get more applicable, better understood, or more correctly applied as you move toward a centralized, one-size-fits-all approach. Against that reality, enter the Food Safety Modernization Act (FSMA). Signed into law in 2011, this sweeping behemoth of legislation directed the U.S. Food and Drug Administration (FDA) to strengthen the food safety system. According to the FDA's website, FSMA "enables FDA to focus more on preventing food safety problems rather than relying primarily on reacting to problems after they occur," which sounds pretty reasonable. As you can imagine though, the devil is in the details, and since its inception FSMA has been a train-wreck of pre-proposed rules, redactions, adjustments, arm-wrestling by special-interest groups, and most importantly: complete and utter confusion for farmers.

As a national-level bureaucracy struggles mightily to find one-size-fits-all solutions for the nation's food safety concerns, farmers are left with very real questions on whether they run a "farm" or a "facility" and whether their direct-marketed food products are subject to the onerous regulations and cost intended for massive multi-national food processing corporations. Most of FSMA is targeted toward fruit and vegetable producers, with the livestock community notably absent from impact. But don't let your guard down, and take the opportunity to advocate on behalf of small-scale regenerative farmers of all types, because you can be certain that our turn is coming. In the meantime, the impacts of FSMA are causing fear and concern among sustainable and regenerative producers to this day, as new proposed rules are generated and issued from on high, in the almighty name of Food Safety.

One-Size-Fits-None

As the FDA searches for one-size-fits-all solutions, the very predicable outcome is that their results look a lot more like one-size-fits-none. Honestly, it's not their fault...put yourself in their shoes for a minute and consider what words you would put into a rule to "fix" food safety at a national level, at all scales, and within all aspects of what is truly a global food system. Personally, I have no idea. What I do know is that localities are always the most in tune with their citizen's needs and are best suited to create legislation that meets those needs. The takeaway for this chapter isn't to harp on regulations or complain from the back pew; instead it is to recognize that the patchwork nature of food safety regulations requires that food producers stay engaged with their local, county, and state regulatory agencies.

You cannot ignore food safety; instead you must become an expert. It's an unfortunate reality that you can't just be smart within your farming operation—you also have to become smart on the multitude of rules that will affect your business. By researching, reading, and educating yourself on applicable food safety regulations, you are also preparing yourself to be an advocate for positive change. If—no, *when*—you find a rule that extends unnecessarily beyond the scope of food safety, promise me that you'll try to change it! Set up a meeting with a regulator, attend those public comment meetings that no one ever goes to, call your mayor or township trustee and tell them how existing policies impact your business. Through advocacy we all can contribute to applying some common sense to a web of regulation that desperately needs it.

Write Your Own Story

Identify regulatory requirements for your proposed enterprise(s), as well as the agencies responsible for enforcement and compliance. Which requirements are non-negotiable and which are…flexible? If you are in the position to, make a list of states and/or counties whose food safety regulations are more amenable to your proposed operation.

CHAPTER 8

Developing Your Digital Customer Base

I know what you are thinking right now: "How on earth am I supposed to develop my customer base if I don't even have a farm yet?" Or something to that effect. I remember my brother-in-law laughing at me when I handed him my newly minted business card, just months after purchasing my land. "Why do you have a business card when you don't have anything to sell?" I may have been a little early on that particular decision, but the truth is that it is never too early to begin identifying, communicating with, and securing potential customers. Keep in mind that the information presented here is not an all-or-nothing approach; some of the concepts you can start to implement now, while others will come later as a natural extension of building your farming business.

For example, I started my website and blog before I had even decided in which state I was going to purchase land! Why would I do that? Because ultimately all of these virtual methods of connecting with people should result in a physical connection someday, in one form or another. Whether that means that someone becomes a customer, an investor, or a cheerleader for you and your farm, your digital presence should encourage or facilitate a physical step toward action on your behalf. Without even knowing it at the time, I was already beginning to apply a concept to my new farming business called the "sales funnel."

The Sales Funnel

Think of this like a funnel, where the wide opening allows a large number of people to connect with your farm and then intentionally guides them toward the narrow opening where action is taken. Not everyone will make it all the way to the action step; in fact, relatively few will. Each step narrows the group until you are left with those who are ready to commit, support, and purchase. You will have some customers who know exactly what they want, know that you provide it, and head straight for full support right off the bat. Most of the time, however, folks need time to consider, compare, adopt, and shift their perspective toward your methods, which can take a lot of time. That's why you need to get people started into your funnel now, because sometime in the near future you'll have products to sell! Don't let yourself get all set up at the farmers market, or arrive at the restaurant with your samples and sales pitch, and have that be the first time anyone has heard of you.

Generally, the steps of a Sales Funnel include (from wide to narrow):

1. Awareness: Help people find you by creating awareness of your operation and endeavors through content sharing
2. Interest: Nurture those interested with more targeted content

3. Consideration: Provide more detailed product information
4. Purchase: Customer buys your product
5. Retention: Keep your customers coming back, create loyalty

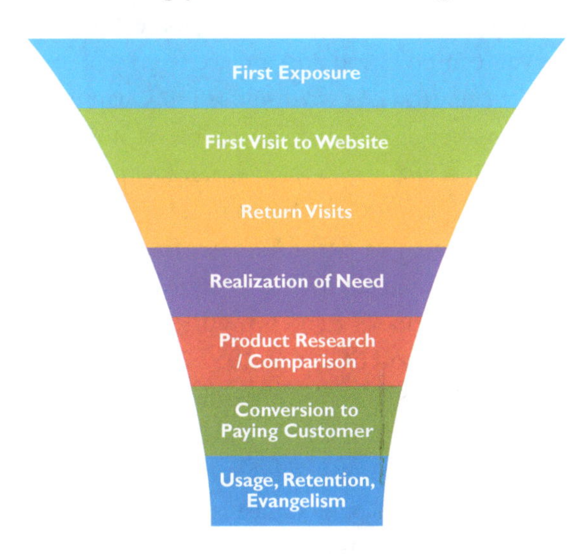

Once your business is fully up and running, understanding this concept will benefit you by providing structure and efficiency in your sales process, helping you understand customer needs so that you deliver the right message at the right time, and showing you where people drop out of the sales funnel so you can continue to refine the process. However, grasping the idea of the Sales Funnel is also valuable outside of the strict application of developing and retaining future customers. The same concept is equally applicable to your broader support base—those who may never purchase but who do:

- Share your social media posts religiously
- Contribute to your crowd-funding loan efforts
- Introduce you to policy makers in your state
- Connect you with career broadening and advocacy opportunities nationwide
- Help you write a book someday

Gathering your supporters, developing awareness of and nurturing interest in your farm, generating action, and creating loyalty —all of these critical concepts are shaped by the different ways, formats, and platforms that you use to communicate. This chapter will concentrate on the digital options that I have used in the past, while more of the physical opportunities will be discussed in later chapters. Combining both aspects will make sure that all of your efforts serve to move someone through the steps of the Sales Funnel toward concrete action on your behalf.

Website

Just as your land provides a physical home for your farm to thrive on, a website provides an online home for your farm—a place where you can be found and a place from which your farm's online presence can grow. As we discussed in Chapter 4, a website is often the first step in creating an online customer experience with your brand and also serves a few other key purposes:

1. Websites help ensure people can find your farm, any time: A 2020 study shared by WebFX shows that 86 percent of people rely on the internet to find local businesses. Moreover, as online food purchases trend upwards (35 million more consumers bought food online in 2019 than in 2018, according to Supermarket News), the importance of having your farm where these consumers search will be increasingly important for your brand awareness and sales growth. In addition, websites are always open, which means people can find and learn more about your operation from anywhere in the world, even when you're fast asleep.
2. Websites serve as an education hub where anyone can learn more about your farm and your products: In fact, most of your activities in digital marketing should drive people back to your website, so they can learn more or even make a purchase. Websites should include essential information that helps establish consumer confidence in your operation—from basic information such as product descriptions, images, and contact information to more sophisticated information, such as descriptions of your management practices and blog posts (more on that topic in a little bit).
3. Websites add credibility to your farm business: By maintaining an up-to-date and informative website you can build credibility for your operation for all who find you online. Conversely, if you do not have a website, customers may perceive that you are not taking this business seriously and this can damage your farm's reputation.
4. A website is the only form of online presence that you own: Your Facebook page, your Instagram account, your Pinterest boards—you don't own those; the platform company does. As a result, you are always at risk of losing your content that lives there if the company decides to change the rules of the platform or if they feel like you have violated any policies.

As I mentioned earlier, I started my website even before I purchased my land, so I kept it simple at the beginning. As I'm sure you could guess by now, I used a free Wordpress design, but you can also pay companies like SquareSpace or others to develop your website for you. In my case, I started with a "landing page," toward which all of the domains I had previously purchased pointed. That main page contained a menu that included "About Us," "Blog Posts," and "Contact Us" tabs. That's it—at the beginning at least. Later on I added pages that detailed my livestock management practices, specific information on breeds I was raising, how to purchase products, and a slew of other information. While a website is necessarily more static than other digital forms of communication, it should by no means be stagnant. At this point in the process, though, all you really need is for someone to be able to type in your website and see something besides a placeholder page.

As you develop your website, keep in mind that every website should eventually include:

1. Easy user navigation: The most important information should be easy to find.
2. Contact information: Make sure this is in an obvious location. Common placement for this is at the bottom of the homepage or in a footer that is on every page. You can also include it as a separate "Contact Us" page and link it from the home page. I also recommend including more than one way for your viewers to contact you (email, phone, or standard contact form).
3. An "About Us" section that has a clear description of who you are: This can include your mission statement, your values, and a little about your farm, team, and/or family.
4. Images and videos: These can help improve the customer experience and help your visitors get to know your operation better.
5. Customer testimonials: These help add credibility and build familiarity for your farm and products and are an important tool in converting consumers from interest to consideration to purchase.
6. Links to your social media channels: Let people who visit your website know what channels they can engage with you on social media.
7. Good SEO: this is a series of strategies designed to help your farm appear higher in search engine results—a critical tool to helping people find you.

Search Engine Optimization (SEO)

The entire point of SEO is to help your website appear in the list of sites that is returned whenever someone types into a search bar. The higher up on the list the better, and the more sites that you own that show up is icing on the cake. In order to accomplish this, SEO is a series of strategies and tactics that earn your website the "head nod" from the search engines whose job it is to legitimize and prioritize the results of a search. When someone searched for "grass-fed beef Ohio," I wanted my website to be the first thing that showed up! There are people who have degrees and full-time jobs pursuing and ensuring optimum SEO for their companies. I say that only to set the stage for the layman's version of what I have to offer on this subject! However, the importance of SEO cannot be ignored and I've learned a few tips and tricks that I will share about slugs, keyword density, and connectivity.

Slugs

When you add pages to your website, they act as sub-pages of sorts, with your main URL being the main page. The identity or location of each subpage is identified with a "slug." For example, my "About Me" page was located at *pasturedprovidence.com/about*, where the extension "about" is the slug. When you build a new webpage, the slug often defaults to "/page1" or something non-descript like that. That is bad for SEO. You want your slugs to be descriptive, using words that you think someone who is searching for you would use. For example, for a Products page, you want that slug to be *pasturedprovidence.com/grass-fed-finished-pastured-beef-ohio* instead of *pasturedprovidence.com/page1*. This type of keyword use in your slugs is an easy first step toward successful SEO.

Keyword Density

Another thing search engines look for when deciding if your webpage is what the search is really looking for is the density of those searched words. To take advantage of this, you need to put yourself in your prospective customer's shoes: what words do you think they will use to search for your farm? Grass-fed? Organic? Non-GMO? Pastured? Location specific? Whatever those words are, they need to be intentionally used in your website text — the more often the better (within reason of course). This keyword density will elevate your site in the list, as the algorithm sees that you offer more of what the search is looking for. Using my "Products" page as an example again, I could just say "I produce beef." Or, I could try my hand at some keyword density, opting instead to say "I raise my grass-fed beef on the lush pastures of southern Ohio, ensuring that they are 100 percent grass-finished without using any grains. This provides a healthy, non-GMO, grass-fed beef product for my customers." The more times you could replace "beef" with "grass-fed beef," the more likely that your site will rise to the top of the list, thanks to your keyword density.

Connectivity

Finally, search engines are looking for legitimacy in their returns. If a website sits out on the interwebs alone and unconnected, it looks like an illegitimate site. However, if your website is connected to a social media platform or two, and has links to a blog post as well, then the search engine is more likely to grant you that sense of legitimacy. This concept is also how you can strive to have multiple returns in the list. That way, when someone searches for "grass-fed beef Ohio," the returns can look like this:

1. *pasturedprovidence.com/grass-fed-finished-pastured-beef-ohio*
2. *facebook.com/pasturedprovidence*
3. *pasturedprovidence.com/blog/the-nitty-gritty-on-grass-finishing-beef-in-ohio*

This is obviously a fictitious example, but it represents what is possible when you pay attention to search engine optimization. Make sure to use appropriate slugs, keyword density, and connectivity to increase your website's SEO, so that when customers search for a farm they find yours!

Blog

"Blog" is one of those words that when you say it over and over (preferably to yourself so no one thinks you're crazy), it stops sounding like a real word anymore. I suppose it actually isn't a real word, in that it is a shortened version of the term "web log," denoting a website that contains written events, stories, and topics. Now that you are equipped to succeed in the next round of Jeopardy with that little tidbit, let's talk about how to use this communication strategy!

Like my website, I started a blog before I left the military and before I even owned the land that I farm. In fact, my initial decision to use Wordpress for my website was partly driven by the fact that Wordpress was originally created as a blog-centric platform with lots of functionality already built in. My blog was meant to be a tool for both tracking my journey into farming, and to generate awareness and consideration about the values I was seeking to build this new operation and chapter of my life around. What I realized a little further down the road was that starting my blog so early in the farm process played a key role in building an audience of interested contacts that would serve my endeavor well in the future. Even though I haven't written a new post in years, to this day I still have people contact me about different blog posts that turn up in their searches. Now that is a powerful addition to my Sales Funnel!

Blogs can be a lot of work, especially if you are not inclined to writing, or find yourself short on time, as many of us do. However, there are several reasons you might consider a blog or similar content channel. The benefits include:

- Consistent communication housed at the center of your farm's online presence: Creating content on one consistent content portal on your website means you have more content to share (on social media for example) in order to attract visitors. In addition, you can bring them to your "home," where you can help guide them through a journey that helps you meet your goals. The alternative is that you share other people's content and send potential visitors elsewhere.
- Establish thought-leadership for you or your farm: A blog is another tool to establish credibility for your farm; further elaborate on your values, practices, and products; and also establish yourself, your team, and your farm as experts in certain areas—whether it be pastured livestock, farmer advocacy, or any number of other things.
- A way to connect with your customers: Building a story about your operation that resonates with blog visitors can go a long way to building a relationship with them, creating trust in your brand and ultimately selling product and creat-

ing advocates for your farm.

- Build your audience: The consistent creation of compelling content, along with the familiarity your blog can breed, provides an increased incentive for customers to opt-in to receiving future email communication from you.
- Search Engine Optimization: According to Hubspot, blogging "helps boost SEO quality by positioning your website as a relevant answer to your customers' questions." Because of the potential SEO benefit and because you want visitors of your blog to also have easy access to other parts of your website, it's important to house your blog on your website (somewhere like *pasturedprovidence.com/blog*) rather than on an independent external blog platform.

Blogs can be pretty simple or can grow in sophistication depending on your time and understanding. Here are some basic best practices that you will want consider if you decide to include a blog on your website:

- Keep your blog name and URL simple.
- Post regularly: This can be a challenge for busy farm team schedules, but making it a priority and using it to further your overall marketing and sales goals can make the time investment worth it. Creating a monthly or annual content calendar where you map out when you want to post and on what can help.
- Consider post length: While there is no consensus on what the ideal blog post length is, it's important that you consider length—and holding your reader's attention—when deciding on the right length. You want your posts to be compelling enough that your visitors feel like they gained some value from reading them and keep them coming back for more. If this can be done concisely, that should be OK.
- Use keywords: Keywords are important because they connect what you are writing to what people search for online. When creating a blog post you'll want to choose a keyword and use it in the content, blog title, and URL for that specific post to help internet searchers find your post. Using the right keywords that are reflective of what your ideal customer is looking for and what you can offer can be really valuable. On the other hand, if you don't choose the right keywords or you aren't specific enough, you may end up attracting an audience that does not truly align or connect to your content and your goals.

- Include a Call To Action (CTA) when possible: CTAs are your opportunity to move visitors through your sales funnel and closer to conversion to whatever goals you have. Blogs are a great way to bring visitors to your website and create more awareness about your farm and products—don't forget to take advantage of this opportunity to transition them into customers while they are there!

When I began my blog it was an effective tool in helping me create awareness about my operation—even before I had an operation. Your blog can be used in the same way by choosing topics to write about that will help visitors become more familiar with you and your farm team, your farm practices, the challenges you face as a farmer, the ways you engage with the community, the products you create and numerous other things. Once you have people at your blog post, you can use different calls to action to help guide your readers to the next step of engagement with your farm. For example, you can ask them to sign up for your email list to hear more from you or direct them to more specific information on the products you sell.

Email

Speaking of email, for many businesses, email marketing is considered one of the best strategies for converting sales. Depending on your goals and how you use email, this can be the case for your farm too. According to a McKinsey & Company study, email marketing is 40 percent more effective at reaching customers than Facebook or Twitter. The same study showed that through email, the buying process happens three times faster than with social media. For this reason alone, it may be well worth it to develop an email marketing strategy. But email marketing isn't just effective—it's a low cost form of promotion to a list that is already qualified and interested in your operation, and it's a great way to build relationships with your customers.

For several years, I managed my email list myself, with a *subscribe@pasturedprovidence.com* email that I then copied and pasted into the Contacts list on my computer. Marketing emails were generated from my normal Mail program and were dull, boring, and text heavy. I also began to run out of room since the Contact lists were limited to 50 names, so I began to have two, then three lists. At this point I realized that I had exceeded my DIY capacity and began looking for an email service provider, meaning a platform that allows people to opt-in (sign up or subscribe) to your emails and unsubscribe. The benefit of this kind of service is that it manages your email marketing list for you, saves you a lot of time, and helps you follow appropriate email marketing regulations (after all, it's illegal to send email campaigns to people who haven't given permission). Some common and affordable email platforms include MailChimp, Hubspot, and Constant Contact. The platforms will all serve the purpose of managing your email marketing list and campaign but will differ in features and price. I now use MailChimp because it is effective and—you guessed it—free!

Once you have a platform, there are several strategies you can use to grow your email marketing list, including:

- Have a sign-up form for your email newsletter on your website, or at the very least an email link where they can subscribe to your list.
- Use your social media channels to let your followers know you have a newsletter they can sign up for and link to the sign-up form.
- Add a checkbox at the point of purchase that asks if your customer would like to sign up for your newsletter.
- If you attend farmers markets, you could also have a physical list available for people to sign up for your emails. I have found that it can be difficult to interpret handwriting, though, and found better success using a laptop to collect addresses. Or if you want to make it even easier, you could have the electronic form on your website available on an iPad for sign up.

So once you have the infrastructure and logistics set up, what should you include in your emails and how often? I quickly learned that, despite the advantages that email marketing brings, simply sending emails does not guarantee success—it's important that your emails serve a purpose for you and, more importantly, provide your readers with value. The last thing you want to do is squander the rare opportunity in which a customer has given you direct access to themselves electronically, which is really what opting-in to an email list implies. What an amazing trust, and one which needs to be treated with respect!

In deciding how often to communicate with your audience, it would be helpful to make a list of all the different things you think you would like to communicate about. You can start with types of emails, such as:

- Product pricing, promotions, and last-chance offers. You could also offer newsletter subscribers special coupons and discounts.
- Recipes, especially those that highlight a product you have available.
- Stories from around the farm, activities from the field, introducing a team member or a farm animal.
- Farm events.
- Sharing relevant farming and food industry news from other sources.

With a little brainstorming, you can come up with a decent list and start to map out topics over a monthly calendar. This activity can help you determine how often you will feel comfortable generating email content and you can weigh that against how much time you can commit to this.

Once you have a plan, be sure to let your subscribers know what they can expect. Setting and meeting expectations and consistency will go a long way in building a relationship with your subscribers. You can also infer that, by signing up for your newsletter, most customers expect there will be some marketing coming their way, but the trick is to ensure your subscribers also get value out of the newsletters. Be intentional about content planning and messaging. In each email you can provide valuable information in the form of education or cost savings, as well as clear calls to action that can help prompt a move down the Sales Funnel.

For example, in one email campaign you can include a story about calving season (with pictures of course!) and what that involves on the farm—fun content, right? Along with that you can include a call to action at the bottom of the email about ordering your beef products or a coupon code that might incentivize purchase. In this email, you are providing entertainment, information, and cost savings. Yes, you're trying to sell, but your readers also gained value from this. In another email campaign, you may share a story about how your eggs are raised and prepared for the farmers market, then a note about your upcoming farmers market schedule so readers know when and where to find you and make a purchase.

The power of email should not be overlooked, but it should never be abused. Provide an opportunity for customers to opt out, and honor that request when they do. Set communication and content expectations, then stick to them. And make it as easy as possible for customers to opt-in to this unique, direct-line, invite-only opportunity. In this way, you'll be ready to maximize your strategic efforts and capitalize on effective email marketing.

As you read this chapter and consider your digital marketing options, don't be surprised if you find yourself uneasy about the idea of treating people like customers. Especially in the beginning, I absolutely struggled with the idea that I needed to constantly sell myself to those around me. Would it be possible to meet someone new and look at them as just a friend instead of a potential customer? How do I tiptoe the line between sharing purchasing opportunities and becoming something I despised, like a pushy used-car salesman or some pyramid scheme purveyor? For me, the question answered itself, as I came to realize that people were genuinely interested in sourcing good food for themselves and their families. All I had to do was put information out there on what I was doing, why I did it that way, and how to get in touch with me, and the rest took care of itself! Don't think of the Sales Funnel as pushing or forcing people towards decisions; instead look at it like you are guiding them toward action by providing them with information that they are already asking for.

Lastly, don't delay. Unlike the legitimate feedback I got from my brother-in-law on my business cards, many of these digital marketing strategies can and should be implemented now, regardless of where you are in your farming journey. Sharing that journey electronically, gathering interested people and getting them moving into your Sales Funnel, can all be started before you actually start farming. That way, when you finally have a product to sell, those digital connections will have resulted in physical ones, as those who choose to become action-oriented customers, investors, cheerleaders, and champions for you and your farm.

Write Your Own Story

Sketch out your website in an outline format, creating a site map for your content. Include potential slugs and a short list of keywords that you are considering.

Make a list of potential customers, along with the best ways to reach them.

CHAPTER 9

The Value of a 'Like' —Social Media Marketing

Social media can be an overwhelming world for many, especially when you mentally take the next step of applying it to your business. Posting pictures of your kids or expressing your feelings on certain subjects is one thing, but how do you make money? Can (and should) you turn "friends" into "customers"? And where exactly should you start? These are just some of the questions that I will tackle in this chapter, but I need you to keep in mind that this world is constantly changing and evolving, which makes it difficult to capture in book form. I'll try to keep things general so that the principles we discuss can be applied to both current and future social media platforms.

Social media is a rapidly changing world that seems to have its own set of rules and even language. Farmers are told that "we need to be on social media," but most of the time that is where the advice stops. Should we be pinning, posting, tweeting, or snapping? And what the heck does any of that even mean? Let's start with a social media overview in order to inform and narrow your focus as you first consider where to post. The statistics I use here are current as of 2021, but feel free to check out *khoros.com/resources/social-media-demographics-guide* for the most up-to-date information. The dominant social media platforms, in no particular order, are:

Facebook

- Social sharing site with 2.7 billion monthly active users
- 96 percent of users access via mobile devices
- Equal parts urban/suburban/rural
- 86 percent of people age 18-29 and 77 percent of people age 30-49 use the platform
- More than 81 percent of all households use the platform, regardless of income level

Instagram (owned by Facebook)

- Social sharing app around sharing pictures and videos
- 1 billion monthly active users
- 51 percent female, 67 percent of people age 18-29
- 83 percent of users say they discover new products and services on Instagram

Pinterest

- Image-driven social discovery site for DIY design, style, beauty, and food
- 322 million monthly active users
- 70 percent female users
- Drives 33 percent more referral traffic to shopping sites than Facebook

Twitter

- Micro-blogging site, limits posts to 280 characters (not including images/videos)
- 330 million monthly active users
- Equal men/women and urban/suburban/rural
- 38 percent of people age 18-29 use the platform

Snapchat

- App for sending videos and pictures
- 381 million monthly active users
- Content is only available for a short timeframe before becoming inaccessible
- Mobile-only platform
- 61 percent female, most used among 15-25 year olds
- Few ways to track metrics or marketing success

LinkedIn

- Business-oriented site to connect career seekers and corporate brands
- 260 million monthly active users
- 60 percent of people ages 25-34 use platform, 57 percent male / 43 percent female
- Typically used for professional services (finance, law, banking, etc.)

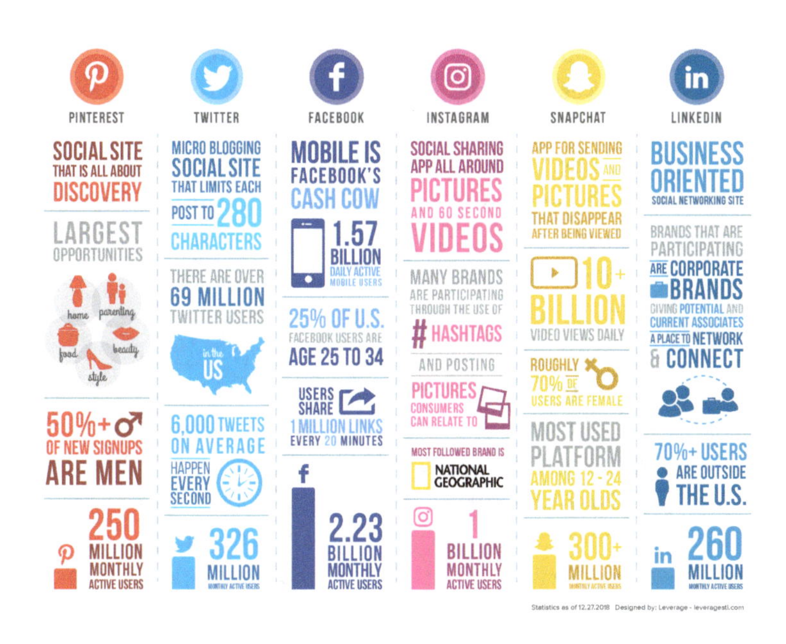

So which social media platform is right for you? The answer, as usual, is "it depends," but here are three important considerations:

1. **Find your customer.** Depending on what statistic you use, 70-80 percent of all consumer purchasing is driven by women, specifically in the age range 30-49. Can you see why that might be important to someone who sells high-end protein options to discerning customers? Of course I sold successfully to both men and Millennials as well, but women with children were my bread-and-butter customers, and I fully anticipate that they will be yours as well. Once you have determined your ideal customer, you'll want to evaluate the data to determine where that ideal customer spends the most time on social media.
2. **Type of content.** The kind of content that is most successful on any given platform, and whether you can provide that type of content, can influence your future success on and decision to use the platform. For example, while most social platforms allow for images, Snapchat in particular is oriented on images that disappear after a short amount of time. Because we want our images to remain in front of our customers, this may not be the platform for you.
3. **Time.** Just like the other digital marketing opportunities, a critical factor when choosing a social media platform is your available time. If you don't have time to create the type of content that performs well on a particular platform, you are not going to be consistent or see results. Consistency is critical to your success with social media marketing, so you need to select a platform that makes sense for your business as well as your ideal customer. Take into consideration the recommended posting frequency by platform as you determine how much time you can devote to sitting in front of your computer, using the following suggestions as a guide:
    - Facebook: 1 to 3 posts per day
    - Instagram: 1 post per day
    - Pinterest: 30 "pins" a day
    - Twitter: 10+ "tweets" a day
    - LinkedIn: 1 to 2 posts per day

There are online tools that can help you save time and be more efficient in your social media marketing (Hootsuite, Tweetdeck, Canva, etc.), but you will also need to invest the time upfront to learn how to maneuver these systems well. In my experience, picking and concentrating on using one or two platforms well is the better option over trying to be everywhere by using a platform management tool.

Given all of that information, my specific social media recommendations (in priority order) for farms and farmers are:

1. Facebook—If you only have time for one platform, make it this one. It is far and away the predominant social media platform, and will be for the foreseeable future.
2. Instagram—Picture oriented is perfect for farms, plus it is easy to dual-post with Facebook since they are owned by the same parent company.
3. Pinterest—Again, picture oriented is good and the user stats line up (70 percent women, median age of 40), but it can be a little more of a chore to learn and use.

Twitter just doesn't lend itself to telling your farming story, and it typically requires too much time to be intentional about tweeting multiple times per day. LinkedIn isn't the target audience, nor does it lend itself to the kind of marketing farmers need. Lastly, Snapchat doesn't create any sort of lasting marketing material, making it a no-go for farmers. While the statistics mentioned here will undoubtedly change overnight, the basic rules for selecting a platform haven't really changed a whole lot over the last few years.

The bottom line is solid: you need to be on social media. So make sure you are choosing a platform that targets your intended consumer, that the type of content matches your capabilities, and that you have the time to devote to posting regularly.

Before we touch on how, when, and what to post, it will be useful to cover some of the unique language associated with social media marketing. The last thing you want is to be "the noob who posts a totes adorbs meme and some troll sends you a NSFW DM," am I right? My tween-aged son helped me write that sentence, and in general I'm astounded by the new language that he and his sisters bring home every day. It really is a chore to keep up with slang, and often I just shake my head and pretend I didn't hear them. I promise by the end of this chapter you'll be able to decipher that sentence, since social media has its own version of slang, but a few terms are foundational to all the platforms and need to be understood:

Handle (@)—A public "username" on a social media account that begins with the "@" symbol. Think of this as a shortcut to your social media presence, so instead of someone searching for "Pastured Providence Farmstead" on Facebook, they just have to type @pasturedprovidence, which is my farm's "handle." You definitely want to have the same handle across all of your social media platforms, in order to make it as easy as possible for your customers to find you. Check each platform you want to use to make sure your desired handle is available before you commit. Ideally, this handle would be the same as, or an abbreviation of, your business name, or website URL. Keep in mind there is a character limit for handles on some platforms, so you want to keep it short and memorable.

Hashtag (#)—A keyword or phrase, spelled out without spaces, with a pound (#) sign in front of it. It is a label for content on social media and is mainly used to denote specific topics of conversation online, allowing someone to search a specific hashtag and view all of the social media posts that have used it in one place. Anything can be a hashtag, but the point is to get noticed by joining a conversation, so you want to use something that others are using as well. A quick Google search will provide a list of the top trending hashtags for each social media platform. You can use a hashtag in the middle of a sentence ("Sometimes this #farmlife gives you lemons; today it gave me lemonade") or you can simply add a bunch of applicable hashtags to the bottom of your post ("Some days are really tough out here, but this amazing sunset makes it all worth it! #farmlife #sunsetsky #adayinthelife").

Engagement—The goal with a post on any social media platform is to attract engagement from your followers via likes, shares, comments, reactions, etc. This is important to understand, now more than ever. Facebook originally displayed newsfeed content in reverse-chronological order, with the most recent posts near the top. If you were "friends" with someone and they posted something, you saw it...plain and simple. Now however, the platform uses an algorithm to organize posts and determine what you see first in your newsfeed. The algorithm uses several factors to determine what, how, and even if content appears. For example, Facebook prioritizes posts from friends and family, posts with more engagement (such as videos with a large number of views or a post with lots of comments), posts with links from trustworthy sources, post types or posts from pages that you engage with often, consistent posting frequency, and posts that reference a trending topic (this is where hashtags come in). The algorithm will also down-rank posts that attempt to force interaction; these are often referred to as "engagement bait." An example of engagement bait would be a post that encourages a follower to "like" or "share" for a chance to win a contest, or one that asks you to "tag a friend" in the comment section. These factors can change at any time, but as of 2021, these are the things to pay attention to. The key point here is that engagement is critical to your posts being seen, but it needs to be organic in nature and not forced.

In order to create engaging posts without falling into the category of engagement bait, I try to use the 50/30/20 rule:

- 50 percent—Engage: On-farm photos, videos, updates, jokes, contests
- 30 percent—Educate: Industry trends and news, professional advice, post shares
- 20 percent—Sell: Products, projects, sales, events

Yes, you're seeing that right...only 20 percent of your social media effort should be specific to sales. I know it's counterintuitive, but customers don't follow brands to be sold things. I will go so far as to suggest, as much as possible, disguising your sales pitch with humor or as a conversation. This is the realm of the soft pitch—not the used-car salesman telling folks that his kids need to eat tonight, so what are they going to buy? Also note, not all the content you share on your social media channel needs to be original content. You can share posts from other brands, experts, and industry leaders, as long as the original source is trustworthy.

Next, let's talk about when to post. If you google that subject, you'll find plenty of advice on what time of day and which day of the week posts are most viewed/clicked/engaged with, but I am going to tell you to ignore all of that advice. What really matters is specific to your target audience—your customers instead of some random online average. You need to determine when your customers are online and what kinds of posts they respond best to. Most platforms offer insight pages showing specific information that will be useful in determining the behavior of your social media followers. Of all the information that is presented within the insights page, your most important data points are the time of day the majority of your followers are online and what types of posts they are responding to. Also be sure to take into consideration your top performing posts, which will provide valuable clues as to what has worked in the past and may be duplicated successfully in the future.

With the how and when questions answered, the obvious progression lands us at what to post on social media. As with everything else in the social media realm, this is a shifting target. What started as short text-heavy sentences about your day, your feelings, or your car for sale has morphed considerably. Platforms like Instagram and Pinterest are built solely on visual content, but even on Facebook you should never post anything without either a picture or a video associated with the post. Why? Because four times as many consumers prefer to watch a video about a product than to read about it. Remember, social media success is all about engagement, and posts with pictures are 39 percent more likely to be shared.

So what on a farm could be so compelling for a customer to want to engage? The short answer: everything! Farms have a massive leg up on other types of businesses when it comes to social media content. Here are just a few ideas that come to mind:

- Behind the scenes, market preparation/setup/tear down, order fulfillment
- Farming tasks and chores; both the mundane and the unique
- Special projects, farm hacks, and on-farm innovations
- Farm updates, decision points, challenges and successes
- Pastoral scenes, rainbows, thunderstorms (and what they mean to farming)
- Recipes showcasing your products
- Farming methods and how they differentiate you from your competition
- Cooperative efforts with other businesses, being sure to link to their handles
- Telling bits and pieces of your story, how you ended up here, and why you care
- Farm bloopers, mistakes, and funny stories
- Introduce your family and employees, how they contribute and work hard
- Examples of customer support for your farm, how it makes a difference
- Last but not least...social media gold: animals (especially babies) and kids!

This list is by no means exhaustive, but I hope you get the point...there is so much social media fodder on a farm that you'll be taking pictures and notes and saving them for a later posting because you've already posted enough for the day/week ahead!

Just like your budget or branding, this portion of your farm business will need organization and planning to be successful. Once you are immersed in your day-to-day operations, it is so easy to push social media content to the side. I recommend using a simple planning template to help you be intentional about posting, keep yourself honest in the busyness of the summer season, and prevent you from missing opportunities to highlight your products (Memorial Day grilling, Veterans Day support and Small Business

Saturday are three specific examples that were especially important for me). It doesn't take long, you can spend a day filling out your planning template while sitting in front of a fire during the dead of winter, as you mentally prepare for the upcoming year. That way, when it is time to start seeds, get grazing, or plant your grains, you'll already have a plan in place to be successful in the social media realm. There will still be plenty of opportune posting moments like when the sunset is stunning, the calving/lambing/kidding starts, a thunderstorm ruins your long hours of making hay, or a customer shares a kind word of encouragement. Those kinds of posts will fill in the gaps, but you'll at least have a plan for what and when to post laid out before you even start your calendar year.

Social media is a powerful tool for telling your farm story.

Even as content is now primarily picture-driven and supported, video is becoming evermore important in the form of "stories" and "live" features. Facebook Live was publicly launched in 2016, and this new method of social media delivery is a direct reflection of what we've already learned—that pictorial and now video content continues to dominate social media engagement. In fact, after Live launched, Facebook found that their users spent three times as much time watching Live videos over traditional recorded videos. A year later, one out of every five videos viewed on the platform Live, generating 35 percent more interaction from the community (likes, shares, etc.) than non-live videos and receiving ten times as many comments.

The general execution of Facebook Live uses a special button within your phone's Facebook app. When you go Live, your page followers receive a notification that you are Live and can choose to tune in for the event. While you stream a Live video, you can see the number of viewers who are watching, as well as a real-time stream of their comments and reactions. After the Live stream is ended, the video is saved to your Facebook timeline just like a normal video. It is this extra level of personal connection that makes Facebook Live so popular and drives follower engagement so high; indeed it could almost be considered a virtual event instead of a post.

In order to fully leverage the power of Facebook Live, here are some tips for success:

- Promote your upcoming broadcast ahead of time
- Ensure you have a reliable signal (or strong WiFi) and a fully charged battery
- Minimize background noise
- Introduce yourself and allow time at the beginning for viewers to join
- Be human, engaging, and interactive
- End your video with a call to action and thank viewers for watching

It's worth noting that Instagram also has a live video feature. Although Facebook owns both platforms, they do differ. As of this writing, Instagram Live allows live videos up to one hour (where Facebook allows 1.5 hours) and Instagram Live videos disappear from your feed after 24 hours (unlike on Facebook, where they are saved to your page like all other posts). That said, by using the Instagram Live feature you can share your live videos with your Facebook audience in real time—and in that way access both audiences at once. Additionally, using the Live function bumps you to the front of the line of Instagram Stories for anyone in your audience, and according to a 2020 Instagram report at least 80 percent of users rely on Instagram to decide whether to buy a product or service. This could serve as a powerful tool to help convert your audience from interested viewers to purchasers.

As with anything, there are certain concepts that just ring true no matter who or why someone is posting on social media. Some of these have been given to me by others and some have been hard-earned lessons from mistakes and situations that I have encountered. The following is my list of best practices, although I would encourage you to take each and every one of them with "a grain of salt." Pass them through your own filter, your own set of marketing goals, and begin to build your own set of "do's and don'ts" with these as a starting point.

Social Media Do's:

1. Be consistent—This is Rule #1 (and probably Rules #2 & #3 as well... it is that important!). You need to stay consistent in your posting frequency, otherwise you fall off the algorithm's radar and your followers' feed. They can't engage with you if they don't see your posts. Use a planning template to build consistency into your schedule, then follow it! Make time in your busy schedule to prioritize social media; otherwise you are wasting your time. Consistency also matters across platforms, so make sure that you keep the same @handle.
2. Use photos, videos, or links—Remember, four times the amount of engagement awaits you! Words are boring.
3. Engage with your followers—Try to respond to, or at least "like," all follower comments on your page. Answer questions, respond to messages, and be present on social media as much after the post as during the actual posting.
4. Be clever, cute, and witty—As much as you are able to, at least...this definitely isn't my strong point. People respond to, engage with and follow those that make them smile. Six Buckets Farm in New Philadelphia, Ohio, is one of the best I've ever seen at this and is worth following just for the daily social media tutorial! For example, instead of posting this picture with a blah caption like "the pigs needed their rest," their caption reads:

"Pig tetris. Need a long one."

Or when posting a picture of two of their little girls climbing their fruit tree, instead of something like bland like "kids will be kids," they chose:

"Six Buckets Farm Branch Managers."

Some of us are better than others in this capacity, but as much as you can, try to be creative and clever in your text and captions. Additionally, in both of these examples you can see how much of an advantage farms and agriculture have over other businesses, especially when it comes to the "social media gold" of children and animals!

5. Tag other pages/people—Using their @handle ups your post's priority in the algorithm and notifies them of your mention, which in turn increases the odds that they comment, like, and/or share said post on their timeline. Together, we all rise.
6. Use #hashtags—When appropriate, but don't overdo it. Use a quick Google search for trending or highly used hashtags on a specific topic, or use a unique one routinely to group your posts under one search umbrella.
7. Follow the 50/30/20 rule—Engage/Educate/Sell, but remember that it needs to be a soft sell, disguised as humor, information sharing, or market differentiation.
8. Keep your page fresh—Update your banner and profile images seasonally, unless you are using your logo as a profile picture.

Social Media Don'ts

1. Post personal stuff—This is a business tool, not a personal page, and must be treated as such. Keep that line drawn—if it doesn't pertain to your business, customer base, or product line, then it has no business in a post. If there is a bill before your statehouse that improves access to local food or supports farmers, then that would squarely apply to your business. Otherwise, avoid politics, religion, and above all...cat memes.
2. Be divisive—It should never be you vs. them. Keep things positive, concentrate on differentiation instead of disparagement, and stay upbeat.
3. Be afraid to share the hard times—Your customer needs to understand that not everything is rosy on the farm... if farming was easy, everyone would be doing it! Limited sharing of trials and struggles, honestly yet positively, can secure good will and drive engagement far more than a beautiful sunset.
4. Use "engagement bait"—Obvious attempts to get comments and reactions actually cost you in the algorithm, so be subtle.
5. Engage in online arguments—No good can come from it. Social media trolls feed on dissension and drama, and the more you argue with them the happier they are.

That final "don't" brings up my final point. If I were to synopsize all of these best practices into a single thought, it would be this: your farm's social media presence is an extension of your brand, which must be protected. If you've gotten to this point, then you have worked really, really hard to create something amazing, both concretely on your soil and virtually online. In the world of social media, there is a feeling that anything goes and people are free to post and comment whatever they want, hiding behind the distance and relative anonymity. My answer is simple: Nope. I don't owe anyone anything. This is my page, my brand, my rules, and what happens on my social media page always remains under my control. No one else gets to take over, any more than a brick-and-mortar business would allow a squatter to take up residence in the middle of their store. There is a balance to achieve here, with conversation, input, questions, and dialog on one side and argument and animosity on the other. I like to think of it like this: I strive for transparency but don't allow trolls.

To help facilitate that balance, I have a One Reply Rule: whenever I post a picture or a comment, anyone gets to disagree or argue their point (respectfully) once. I always get the last word, either to respond to their argument, provide a more detailed explanation of my viewpoint or simply acknowledge their perspective. Any other follow-up dissent or disagreement comments from that person get deleted—period. They get one reply and that is it. That might seem harsh to some, and I can almost hear the trolls now…"What about freedom of speech?" Elsewhere, sure. On my page, as my business, under the umbrella of my brand—absolutely not! You have to protect your brand, especially in the world of social media where people seem to think that anything goes. My page, my brand, my rules.

Knowing the benefits and limitations of each social media platform and how they apply to your business is just the tip of the iceberg. To be successful in social media marketing, you have to talk the talk before you can walk the walk; otherwise you risk being "the new guy who posts a totally adorable humorous image and some disagreeable person sends you an R-rated not-safe-for-work direct message" (see, I promised you could read that by the end of the chapter!). Ultimately, success in social media is all about getting your audience to engage with your content, and the 50/30/20 rule provides some guidance about how to secure that. Make use of your platform's insight analytics page, disguise your soft sales pitches with humor or empathy, avoid the "engagement bait" trap, and use a planning template to map out your social media year while things are relatively slow. All of these tips will help keep you on task in the midst of the busy season, ensuring that your posts reach your intended customer base.

Unfortunately, this chapter is pretty much doomed to be outdated as soon it gets published, so concentrate less on the details of the platforms and more on the foundational concepts. Facebook and Instagram Live represent the newest iteration of the ever-changing social media world, and are examples of the importance of visual content over the historically textual content. Luckily for us, farms provide some of the most exciting, compelling, heart-wrenching, and fulfilling images of any business out there…it really isn't fair how easy we have it when it comes to social media content!

Most importantly, whatever content you put out there, you must protect it as an aspect of your farm brand. Using these concepts will tame the crazy, intimidating, wild wild west of social media, allowing you to harness the value of a "like" for the benefit of your farming business.

Write Your Own Story

Research and list some possible handles, making sure they are available across all your desired social media platforms. For every month of the year, identify natural posting opportunities (events, holidays, etc) and begin to plan out your social media marketing calendar.

CHAPTER 10

Consumers Want to Connect with Their Food

Whenever I have the opportunity to talk to a group of consumers instead of farmers (although none of us really escapes the fact that we all consume food every day), my message is one of value. I'm afraid that unless we as a community of eaters actually value food again, no lasting progress in addressing food system issues will be realized. Especially as Americans, we lack a strong food culture that places the importance of food where it belongs: above technology, entertainment, or travel. This isn't an indictment for any specific individual as much as it is an acknowledgement of what really needs to change before meaningful progress can be made in fixing the broken food system in this country.

I hesitate to judge individuals for this mis-prioritization, in part because food is one of the most confusing environments for a consumer to navigate. For something that is so critical to our survival, we sure do have a lot of questions about how to do it right. It is a topic where the intensity of opinions rival religion, misinformation and deceit are routine, and the seemingly endless claims made by the multitude of companies involved in the supply chain muddy the waters instead of providing clarity. That father of three frantically scanning the brands for some sense of peace in his decision, or that single mother stocking the pantry for her children while quietly questioning her choices...they both want to do right. In fact, I've come to believe and trust that we all have our hearts in the right place and at our core want the best for us and ours. So how is it that we as a people share so little consensus about what defines quality food and how to make good food choices?

The biggest problem in food is that our customer doesn't know who they can trust.

Consumer Confusion in Food

There are so many different messages being sent to consumers today about what they should and shouldn't eat—by food companies, grocery stores, and, yes, even farmers—that the end result is mass confusion. It is against this backdrop of uncertainty that you are entering the fray to produce food for others. I don't say all this to scare you, but instead present this reality as an opportunity—dare I say, a mandate—for a different approach. If you are to be successful in reaching this overtasked and oversaturated audience with your product, you will need to communicate the value intrinsic within your food. There are several options to do this, including third-party certifications. But too often farmers rush to the certification route without considering another, potentially more powerful option: transparency. There is a place for both, which we will discuss at length throughout this chapter. In general, I am in support of certain labels and I personally purchase products based on them; however, I also successfully built my farm without being certified anything, instead leveraging the power of transparency.

Back in 2013, a news story was released but immediately buried under the gossip surrounding the Tony Awards and the latest PR nightmare for our government caught spying on its citizens and lying to Congress. An outbreak of Hepatitis A was linked to organic frozen berries, offered through Townsend Farms of Fairview, Oregon. How could this happen? Weren't certified organic products supposed to be better, healthier, and free from contaminants? After reading the article, the bigger picture began to take shape, and among other things it revealed what I view as the biggest chink in organic food's armor.

Organic food offers consumers an improvement over "conventional" agriculture, in that pesticides, herbicides, fungicides, antibiotics, and synthetic chemicals normally applied carte blanche in heavy doses are not allowed. Animals are given more space and "access" to the outdoors than normally afforded their conventional counterparts, who are packed in cages, pens, and stalls tight enough to walk across their backs without ever touching the ground. However, when people buy organic food, they often have a mental image of what they think they are supporting that doesn't necessarily reflect reality. While most "normal" chemicals are not allowed, plenty of them are. Industrial producers have taken the definitions and rules and pushed them to the very brink.

Ultimately, certified organic is an absolute improvement over conventional agriculture, but in many cases it falls short of ideal by allowing consumers to continue buying their food from a faceless source that advertises falsehoods as truth and keeps the veil in place between people and their food. Simply put, it is extremely difficult for quality food to come from an industrial model no matter what certification is attached. While the organic food movement started with a vision of small farm producers doing things the right way—valuing soil health, eschewing synthetic and unnatural chemicals, and striving for humane animal treatment—it has now largely become the stomping grounds of Big Ag. In that process, organic has lost its greatest strength: its moral foundation. When the industrial system got its hands on organic, the label (and the product) quickly became adulterated and diluted by the "bottom line," the inclination to cheat and cut corners to save a buck, and the impulse to weaken standards in order to fit organic standards into an industrial mold.

In the news story I mentioned above, the problem is identified in these statements: "The fruit came from around the world: Chile, Argentina, and Turkey. The pomegranate seeds processed in Turkey appear to be the culprit." Unfortunately, the international origins of the fruit blend was hidden from consumers, replaced by the bucolic image of Townsend Farms and its warm and fuzzy tag line: "Since 1906, Field to Farm to Family." Again, the core issue here is another blow to that all-important facet of any relationship —trust. Take a look at the chart on the next page from the Cornucopia Institute, showing the parent companies who have purchased organic facets as of 2016.

As you can see, more big industrial companies are trying to get in on the action and others are increasing their organic holdings. Industrial agriculture involves a faceless producer who employs a fleet of lawyers and has little to no stake in the "goodness" of their product—only a drive to adulterate their food-like offerings to the point where it no longer resembles its noble origins. When the organic food movement was gobbled up by Big Ag, it fell into the same trap, and every so often the curtain gets pulled back to reveal the ugly truth. Recognizing the reality that confusion and misinformation is the norm in the world of food production is the first step in positioning your farm's alternative message for success.

Labels as an Economic Decision

There are many in the world of "sustainable agriculture" who believe in the sanctity of the certified label, especially the organic label. You are "less than" if you aren't certified—missing out on an amazing opportunity at best, and considered an outsider and a charlatan at worst. Some of this fervor is understandable, especially from the old guard, when you consider what they went through to buck the chemically intensive industrial system and pave their own way to a holistic, regenerative approach that ultimately became the USDA Organic program. However, they take it too far to conclude that if you aren't certified then you are simply unable to produce clean, unadulterated, humane, and ecologically sound food. That is patently false. I know, because I did it for the last eight years without the assistance of any certifications, thank you very much.

Simply put, labels are a choice, and in some situations a very good one. But they are by no means a requirement to do business in this world of agriculture. However, if you believe so strongly in a specific label that participation for you is not voluntary (harkening

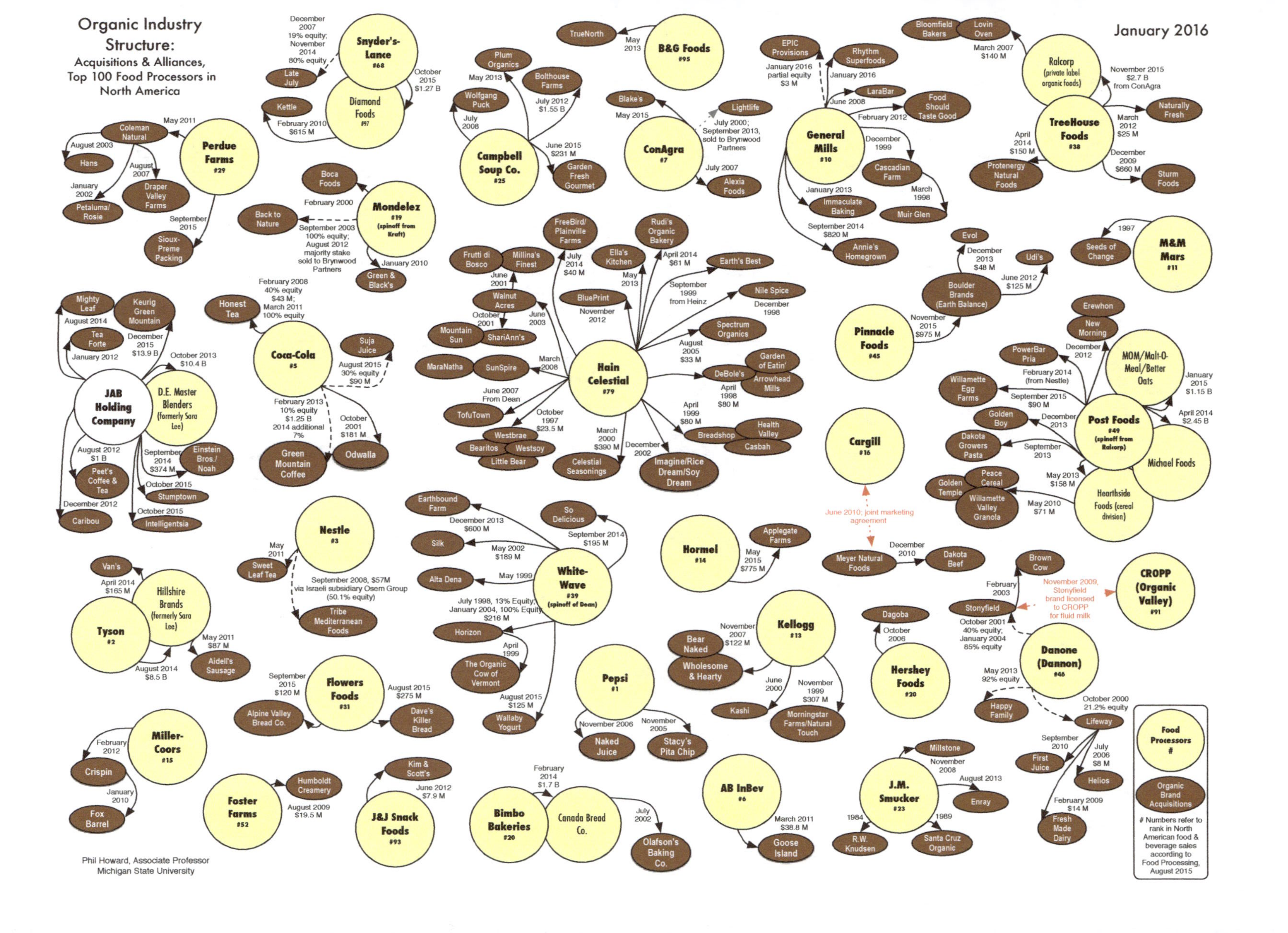

Organic Industry Structure:
Acquisitions & Alliances, Top 100 Food Processors in North America
January 2016
Food Processors #
Organic Brand Acquisitions
Numbers refer to rank in North American food & beverage sales according to Food Processing, August 2015
Phil Howard, Associate Professor
Michigan State University
Perdue Farms #29
Coleman Natural
May 2011
Hans
August 2003
Draper Valley Farms
August 2007
Petaluma/ Rosie
January 2002
Sioux-Preme Packing
September 2015
JAB Holding Company
D.E. Master Blenders (formerly Sara Lee)
October 2013 $10.4 B
Keurig Green Mountain
December 2015 $13.9 B
Mighty Leaf
August 2014
Tea Forte
January 2012
Einstein Bros./ Noah
September 2014 $374 M
Stumptown
October 2015
Intelligentsia
October 2015
Peet's Coffee & Tea
August 2012 $1 B
Caribou
December 2012
Snyder's-Lance #68
Diamond Foods #97
October 2015 $1.27 B
Late July
December 2007 19% equity; November 2014 80% equity
Kettle
February 2010 $615 M
Mondelez #19 (spinoff from Kraft)
Boca Foods
February 2000
Green & Black's
January 2010
Back to Nature
September 2003 100% equity; August 2012 majority stake sold to Brynwood Partners
Coca-Cola #5
Suja Juice
August 2015 30% equity $90 M
Odwalla
October 2001 $181 M
Green Mountain Coffee
February 2013 10% equity $1.25 B 2014 additional 7%
Honest Tea
February 2008 40% equity $43 M; March 2011 100% equity
Hillshire Brands (formerly Sara Lee)
Tyson #2
August 2014 $8.5 B
Aidell's Sausage
May 2011 $87 M
Van's
April 2014 $165 M
Nestle #3
Tribe Mediterranean Foods
September 2008, $57M via Israeli subsidiary Osem Group (50.1% equity)
Sweet Leaf Tea
May 2011
Flowers Foods #31
Dave's Killer Bread
August 2015 $275 M
Alpine Valley Bread Co.
September 2015 $120 M
Miller-Coors #15
Crispin
February 2012
Fox Barrel
January 2010
Foster Farms #52
Humboldt Creamery
August 2009 $19.5 M
J&J Snack Foods #93
Kim & Scott's
June 2012 $7.9 M
B&G Foods #95
TrueNorth
May 2013
ConAgra #7
Lightlife
July 2000; September 2013, sold to Brynwood Partners
Alexia Foods
July 2007
Blake's
May 2015
Campbell Soup Co. #25
Bolthouse Farms
July 2012 $1.55 B
Garden Fresh Gourmet
June 2015 $231 M
Plum Organics
May 2013
Wolfgang Puck
July 2008
Hain Celestial #79
Earth's Best
Nile Spice
December 1998
Spectrum Organics
August 2005 $33 M
Garden of Eatin'
Arrowhead Mills
DeBole's
April 1998 $80 M
Health Valley
Casbah
Breadshop
April 1999 $80 M
Imagine/Rice Dream/Soy Dream
December 2002
Rudi's Organic Bakery
April 2014 $61 M
September 1999 from Heinz
Ella's Kitchen
May 2013
BluePrint
November 2012
Celestial Seasonings
March 2000 $390 M
FreeBird/ Plainville Farms
July 2014 $40 M
Millina's Finest
Walnut Acres
June 2003
Frutti di Bosco
June 2001
ShariAnn's
Mountain Sun
October 2001
SunSpire
March 2008
MaraNatha
TofuTown
June 2007 From Dean
Westbrae
Westsoy
Bearitos
Little Bear
October 1997 $23.5 M
General Mills #10
Rhythm Superfoods
January 2016
LaraBar
June 2008
Food Should Taste Good
February 2012
Cascadian Farm
December 1999
Muir Glen
March 1998
Immaculate Baking
January 2013
Annie's Homegrown
September 2014 $820 M
EPIC Provisions
January 2016 partial equity $3 M
TreeHouse Foods #38
Ralcorp (private label organic foods)
November 2015 $2.7 B from ConAgra
Naturally Fresh
March 2012 $25 M
Sturm Foods
December 2009 $660 M
Protenergy Natural Foods
April 2014 $150 M
Lovin Oven
Bloomfield Bakers
March 2007 $140 M
M&M Mars #11
Seeds of Change
1997
Pinnacle Foods #45
Boulder Brands (Earth Balance)
November 2015 $975 M
Udi's
June 2012 $125 M
Evol
December 2013 $48 M
Post Foods #49 (spinoff from Ralcorp)
MOM/Malt-O-Meal/Better Oats
January 2015 $1.15 B
Michael Foods
April 2014 $2.45 B
Hearthside Foods (cereal division)
Erewhon
New Morning
December 2012
PowerBar Pria
February 2014 (from Nestle)
Willamette Egg Farms
September 2015 $90 M
Golden Boy
December 2013
Dakota Growers Pasta
September 2013
Peace Cereal
Golden Temple
Willamette Valley Granola
May 2013 $158 M
May 2010 $71 M
Cargill #16
Meyer Natural Foods
June 2010; joint marketing agreement
Dakota Beef
December 2010
CROPP (Organic Valley) #91
November 2009, Stonyfield brand licensed to CROPP for fluid milk
Danone (Dannon) #46
Stonyfield
Brown Cow
February 2003
October 2001 40% equity; January 2004 85% equity
Happy Family
May 2013 92% equity
Lifeway
October 2000 21.2% equity
Helios
July 2006 $8 M
First Juice
September 2010
Fresh Made Dairy
February 2009 $14 M
Hershey Foods #20
Dagoba
October 2006
J.M. Smucker #23
Millstone
November 2008
Enray
August 2013
Santa Cruz Organic
1989
R.W. Knudsen
1984
Kellogg #13
Morningstar Farms/Natural Touch
November 1999 $307 M
Kashi
June 2000
Bear Naked
Wholesome & Hearty
November 2007 $122 M
Hormel #14
Applegate Farms
May 2015 $775 M
AB InBev #6
Goose Island
March 2011 $38.8 M
Pepsi #1
Stacy's Pita Chip
November 2005
Naked Juice
November 2006
Canada Bread Co.
Olafson's Baking Co.
July 2002
Bimbo Bakeries #20
February 2014 $1.7 B
White-Wave #39 (spinoff of Dean)
So Delicious
September 2014 $195 M
Earthbound Farm
December 2013 $600 M
Silk
May 2002 $189 M
Alta Dena
May 1999
Horizon
July 1998, 13% Equity; January 2004, 100% Equity $216 M
The Organic Cow of Vermont
April 1999
Wallaby Yogurt
August 2015 $125 M

back to our values conversation from earlier), then by all means strive to make that work knowing from the get-go that you will be working within the constraints of that certification. From my perspective, I believe that any certification should start as an economic choice. If it doesn't pay for itself or help you sell product, then why bother?

Let's start with a conversation about need, before we even tackle the financial accounting, by considering a label's purpose: it is intended to communicate value and differentiate the product it adorns from those around it on the supermarket shelf. Inherently, it is meant for a nameless, faceless product that has to jump out at that single mother or father of three I talked about earlier, in the ten or so seconds that they are standing there agonizing over their food decisions.

If you choose to sell directly to consumers, especially face-to-face at a market, buyers club, or restaurants, then I would argue that you probably don't need that label; you have a more powerful weapon in your marketing arsenal: you and your story! In my experience, a connection with the farmer trumps a certification label every time. When a customer sees the passion in a farmer's eyes, hears about the way that they run their operation or, even better, sees it in person, that connection is incredibly powerful. Would being certified help bring people to your website if they are searching for "organic carrots" or "grass-fed beef"? Perhaps, but using the techniques you learned in Chapter 8 will help address those key words. The bottom line is that in a direct-selling situation, there are some scenarios where a certification might help you, but in no way is it a must.

That reality changes significantly if you decide you want to pursue a wholesale avenue, especially if that puts your product on a shelf or in a freezer somewhere where you aren't there to speak for it—that is when a label begins to make far more sense. Using a certification as one of your new cooperative's production standards might be a smart move since you don't have to recreate the wheel; as long as the shoe fits, go ahead and wear it. If you plan on selling into a local grocery store or through a distributer, then absolutely weigh your options for certification and see if the math pencils out. The first step in the thought process should be whether you need it or not. If the answer is "maybe," then it is time to look at the logistical aspects to make sure they are in place.

The overall concept of any certification involves certain production standards, but most also have additional requirements for transportation, storage, and packaging—in other words, your entire supply chain must also be compliant. For example, prior to harvesting organic corn you have to fully clean out your combine header and collection bin to remove any non-organic grain and residue. You even have to write-off a certain amount of your organic crop as non-organic so that it serves to "purge" the entire system. The same goes for trucks that carry it and grain bins that store it. Take that same type of regimentation and then turn it up a notch if live animals are involved!

Organic livestock (with the notable exception of dairy, which changed the rules for itself thanks to a powerful lobbying arm) have to be born from already certified mothers, raised on organic ground, fed organic feed and pasture, and then (most importantly to this logistics conversation) processed and packaged at a certified organic facility. This is another reason I never became certified organic. As of this writing, there is exactly one organic processor in the entire state of Ohio—a true tragedy. So even if I did have a need, and decided that it did make financial sense, I still couldn't make anything happen without the logistical piece in place.

Financially, transitioning my land to organic and adding an organic livestock "scope" (another layer of certification independent of the land itself) would cost me $1,400 per year, even before I paid a premium price to a theoretical organic processor. At present, there are cost-share dollars available through the Farm Bill and administered through the Farm Service Agency, up to 50 percent annually, but that money is in no way guaranteed. To that point, the 50 percent cost share for 2021 was a shock to everyone, as previous years producers had enjoyed a 75 percent cost-share reimbursement! Using my grass-finished ground beef as an example, which I sold at $10 per pound: there is little to no space available above my existing price to be able to recoup those direct fees, not to mention all of the additional paperwork and time associated with tracking and logging animals to USDA's satisfaction.

As you can see, it isn't as simple as "you need to be XYZ certified in order to be profitable." In fact, sometimes certification can represent a labor of love instead of a sound financial choice. That said, if you do choose to be certified, you should do your research and select the certification that will give you the most bang for your buck and that will be most beneficial to your marketing outlets. By definition, the certification needs to be recognizable to a large audience, so most of the time I would recommend one of the following (in no particular order) as good options to start with:

1. Non-GMO Project Verified

It is my experience that the American consumer is more concerned with sourcing non-GMO products than any other "qualifier" in food. This concern is so strong that more often than not, certified organic products still feel the need to highlight that they are also non-GMO, despite the fact that GMO ingredients are not allowed in organic production. The Non-GMO Project is a non-profit organization that provides the labeling program for products grown without using genetic engineering, verifying that the process the products go through "from seed to shelf" are produced according to their rigorous best practices for GMO avoidance.

2. USDA Organic

Run by the USDA's National Organic Program, this label requires products to contain at least 95 percent organic ingredients and prohibits synthetic growth hormones, antibiotics, pesticides, biotechnology, synthetic ingredients, or irradiation used in production or processing. Although it is advised and guided by a National Organic Standards Board of farmers and scientists, organic certification has continuously dealt with the issue of being subject to government bureaucracy and the influence of large, well-funded, multi-national corporations seeking to water down the original intent of the program. However, it is still the most recognizable and prevalent food label available to differentiate products to consumers.

3. Certified Naturally Grown

Essentially, Certified Naturally Grown follows the same standards as USDA organic, but without being subject to the high cost and bureaucratic process associated with organic. This alternative is administered by a non-profit organization, with other certified farmers acting as inspectors to verify compliance. The program was designed with smaller operations in mind—with the belief that the National Organic Program is designed more for medium and large-scale producers. The paperwork for Certified Naturally Grown and the certification dues are kept more affordable for small-scale producers. Keep in mind that there is no assistance within the Farm

Bill for CNG certification, unlike USDA Organic, which still has some cost-share help available.

4. Animal Welfare Approved

Humane animal treatment is also a key consideration for consumers sourcing meat, eggs, and milk. Animal Welfare Approved created standards that cover the way its participating farms raise their animals (including beef and dairy cattle, bison, sheep, goats, pigs, chickens, turkeys, ducks, geese, and rabbits), stating that the basic premise of their standards is that animals must be able to behave naturally and be in a state of physical and psychological well-being. They only certify family farms, charge no fees to participating farmers, and require that animals must be raised on pasture or range.

5. AGA Grassfed

The American Grassfed Association certifies ruminant animals (cattle, bison, sheep, and goats) that are fed only on pasture, in addition to being raised without antibiotics, synthetic hormones, or confinement and with standards for high animal welfare. Unlike USDA Organic or Certified Naturally Grown, there is not a certified processor requirement to keep the label chain intact, keeping this label as an option for livestock ranchers who do not have access to a certified organic / CNG processing option.

These are just the major label players, which I would recommend sticking with if you decided to pursue a third-party certification for your farm. There are others, including a growing Regenerative Organic Certification, but you need to stick with consumer recognition in order to reap the full benefit of paying for the certification. Again, start with need first and don't let other folks who are more religiously supportive of any of these labels spend your hard earned money for you. Then make sure that the full "label chain" logistics are in place to see your product through to the shelf or store cooler with the label you paid for intact. Finally, be sure to weigh the full financial costs for any of the certifications, making sure that they pay for themselves. Ultimately, matching your marketing situation with the label that will help you stand out from the crowd will be your best choice, but always keep in mind that it is just that—a choice and not a requirement.

The Power of Transparency and Your Story

Despite the fact that I shop and buy organic as much as possible in the grocery store, it didn't make sense for me to pursue certification. There wasn't a need, a legitimate logistical pathway, or a financial incentive that justified the costs. Instead, I pursued what I will call "relational marketing," in which I pursued, developed, and capitalized on a relationship with a customer that is built on trust. I got a premium price for my products for one reason: people believed in me, my production methods, and my story. In short, I was able to successfully leverage the power of transparency.

Transparency was one of the cornerstone concepts of my farm, and it is completely foreign to the industrial model of agriculture. You'd need an appointment, a badge and a space suit in order to visit the poultry barn where all those chicken nuggets are raised. Why? Because if someone saw/heard/smelled what went on behind those closed doors, they'd be jumping on the vegan bandwagon as fast as their pocketbook could take them. Industrial food production is disgusting, filthy, and inhumane. But take heart, American eater: farms like ours represent another option, one where customers can better connect with and value the food they eat.

The first step in pursuing transparency is to share your journey, your decision making, and both your successes and failures. This creates a sense of community, promotes good will, and elicits the emotional connection that is so lacking in nameless, faceless food. This might sound calculated, and to a certain degree I suppose it is, but it certainly isn't disingenuous. When I created a blog post describing the pressure of trying to get the hay baled while storm clouds threatened, and the intense disappointment when the raindrops started falling before I completed the field—that is real. When I posted pictures on Facebook of a cow down with a displaced uterus, the struggle to save the calf, and the emotional victory when it stood and walked—that is real. Be real with your customer base and they will love you for it. Let them in to your life as a farmer and you will establish a relational bond that is more powerful and more profitable than any label could possibly be. Be mostly positive, but don't be afraid to share struggles as well—it is in those moments that you become their farmer raising their food, even before you show up at the farmers market or at their door for a delivery.

This approach is a two-way street, though. When you open yourself up to the general public, you also open yourself up for questions, challenges, and disagreement. Generally, I consider that a good thing and encouraged my friends, followers, and customers to challenge me if they saw something that seemed incongruous with my beliefs. I'm pretty sure that military pilots invented the concept of "constructive criticism," and one didn't make it very far in my previous career without having thick skin! The best thing that can happen is a simple response that clears the air. But maybe something is revealed that I either overlooked or didn't consider, and we all should be willing to take that input to heart. You will need to set your own boundaries based on your comfort level with "letting them in" that is healthy and safe and mostly common sense.

A while back, someone who follows my Facebook page commented on a picture I had posted of my new pigs. In the picture, she noticed that the pigs were in a small enclosure of some sort, and she said "Just curious, but why are they in a small pen?" She was implying that, in her perception, the way these pigs were being housed might not be compatible with my professed values, and requested an explanation. Everything was just fine—they happened to be enclosed in my stock trailer for the picture, awaiting release into their outdoor paddock. To some, this might have been a little forward of this customer, or somehow out of place. To many ranchers it likely would have ruffled some feathers in a "none of your business" sort of way. But this customer of mine knew that she could safely ask that of me without fear of negative blowback, because one of my core values was transparency. And because I didn't take offense and instead shared with her (and everyone else who may have been watching the exchange from the sidelines) what was actually happening, her relational bond with her farmer was strengthened.

The concept of transparency doesn't just cover digital media and marketing; it includes in-person interactions as well. For folks to see a real, working farm is a treat these days, even in rural areas where you might assume that everyone is already well-versed. Even if they are familiar with agriculture in general, it still pays to show off what you are doing on your land. Their connection is what you are cultivating—with your operation, with your farm, with your story, with you. I've even gone so far as to invite customers to attend and participate in poultry processing days, which would land more toward the extreme end of the transparency spectrum! I had very few people accept the invite, but the ones who attended left with a greater sense of appreciation for their food and for my efforts to provide it—never a bad thing. One way to prepare the ground for these interactions involves an old-school job-seeker tool called the "elevator speech."

Elevator Speech

Imagine yourself approaching a chef, grocery store manager, or potential customer out of the blue. You have one chance to quickly, accurately, and compellingly present yourself and your farm before they move on to the next thing demanding their attention. You approach them, smile, shake their hand, look them in the eye, and then…what would you say? Is this a situation where you really want to "wing it" or stumble over your words? Of course not, which is why you need to develop your elevator speech.

The Small Farm Program at the University of California, Davis defines the elevator speech as "a clear, brief message or 'commercial' about you. It communicates who you are, what you're looking for and how you can benefit a company or organization. It's typically about thirty seconds—the time it takes people to ride from the top to the bottom of a building in an elevator. (The idea behind having an elevator speech is that you are prepared to share this information with anyone, at anytime, even in an elevator.)" For our purposes, the goal is to be able to communicate the critical and compelling aspects of our farming story to a potential customer in a short, well-rehearsed, natural-sounding speech.

Most of the advice or guidance you'll find online on this topic revolves around traditional job seekers, but is easily adapted to our agricultural purposes. For example, the general outline that UC Davis recommends remains the same:

- About you: who you are and what you do
- What you offer: what problems are being solved or contributions made
- What are the benefits: very special product, service, or solution that you offer
- How you do it: concrete example that shows uniqueness

In addition to the general outline, here are some other tips to help you create your elevator speech:

- Be brief: no more than thirty seconds, or approximately eight to ten sentences
- Be persuasive: strive to spark the listener's interest in your products
- Practice makes perfect: you want a natural-sounding conversation versus an obviously rehearsed speech
- Be positive: elevate your methods, avoid denigrating others
- Don't rush: while you want to convey a lot of information, don't do it by speaking quickly
- Be flexible: Seldom will your speech be the same every time, so concentrate on nailing the foundational pieces and then let the situation and audience determine the details

Don't let all of that scare you; this is really a very simple concept that you likely already have a 90 percent solution for. The key pieces are the brevity, simplicity, and practice that will let your pre-prepared speech sound natural and flow easily. As an example, here is my elevator speech for Pastured Providence Farmstead:

> Hi there, my name is Paul Dorrance, and I own and operate Pastured Providence Farmstead, a pasture-based livestock operation in Chillicothe, Ohio. I raise 100 percent grass-fed and finished beef and lamb, as well as pastured non-GMO pork, poultry, and eggs. All of my animals are raised on pasture, enjoying fresh air and sunlight their entire lives, which creates amazing-tasting, healthy, and humanely raised meat and eggs. If you are interested in learning more about my operation or my products just let me know—I love sharing all of the cool things that are happening out on my farm!

My speech covers the basics of who I am and what I offer, but also touches on how and why I do it as well as the benefits of my methods in a positive manner. It takes me twenty-six seconds to say, and it leaves the audience with an open door to pursue more in-depth conversations moving forward.

The concept of an elevator speech might seem intimidating or unnecessary, but I promise you that neither of those are true. If you are going to be successful selling your farm products, you'll almost certainly achieve that by talking to someone, and you'll want your elevator speech prepped and ready to help you do that. There will be countless opportunities to address potential customers, whether you go to them wherever they are or invite them to come to you. Either way, you'll find yourself giving a version of your elevator speech more often than you could imagine.

Go to Your Customer

I know you are probably saying, "I got into farming so that I didn't have to talk to other people," but that isn't going to help sell your products. Especially when you are just beginning, you don't ever want to turn down an opportunity to share your story with others, talk about your available products, and collect contact information for those who indicate interest. As I was getting started, there were almost countless chances to connect with potential customers, mostly because I made myself available to go to them.

My first opportunity to speak to an audience about my farm was the "Chillicothe Garden Club." Sounds nice, doesn't it? A new customer I met at the local farmers market invited me to speak to her club's monthly meeting about the importance of food, and I jumped at the opportunity. Turns out, the self-proclaimed Garden Club was a handful of ladies who got together, drank wine, gossiped about other ladies, and ogled the (relatively) young man standing in front of them in the cowboy hat for an hour. It felt like a complete waste of time and was more than a little uncomfortable!

A short time later, my insurance agent invited me to speak to the local Rotary Club, where he was a member. At least that was a little more of a professional crowd, who seemed legitimately interested in what I had to say instead of what I was wearing. I gave it my best shot, espousing the virtues of pasture-based animal agriculture and hailing the many benefits to soil, animal, and human health. Folks were receptive, and asked questions, but at the end of my time I left without any concrete evidence of success. It was enjoyable at face value, but disappointing in the lack of immediate expansion of my customer base.

Fast forward three years and I get two phone calls out of the blue, just several weeks apart from each other. One was from a woman who grabbed my card at the Chillicothe Garden Club fiasco who was interested in purchasing a half-hog. The other call was from, you guessed it: a member of the Rotary Club who had heard me speak. As it turns out, my talk was the first time he had been challenged to think differently about food and he had been on a slow yet steady journey to wellness ever since. He became a loyal customer of mine and regularly supported the farm directly as well as connecting me with others in the local community.

Most of the time a response doesn't take that long to reveal itself, but I tell those two stories to illustrate this point: sometimes you just don't know in the moment whether or not an opportunity will be fruitful. Sometimes you can tell right away, others will never result in anything, but more often than not I've found that putting myself, my farm, and my story in front of local groups and individuals is worth my time and effort. Seek out your local Rotary, book clubs, VFWs, American Legions, and Lions Clubs and ask them if a presentation would be something they are interested in. Often, word will get around that you have something different to say and those groups will come find you.

A slightly different way that I reached out and connected with others was through the local Chamber of Commerce. They were stunned when I came in to discuss my business and become a member. I was only the second farm or agricultural business to ever participate in the Chamber! While their traditional role of facilitating business relationships wasn't generally as useful to me, I was able to attend and even host events like a "Business after Business" cocktail hour or a "Professional Speed Dating" awareness session. When I began to host my farm school it was the Chamber who had a conference room available for its members, an amazing resource that more than paid for my annual membership by itself. While you may not have a downtown storefront in which to hang your Chamber of Commerce sign, don't minimize the importance and value represented in participating in your local economy and business activities.

Meeting local business owners and entrepreneurs not only helps raise their awareness of your presence, but it opens up opportunities to give back to your community (and gain the social capital that comes with it). One of my Chamber connections reached out to me a few years ago, as a group of citizen activists were trying to raise money to design and build a children's museum. The Mighty Children's Museum was a community-wide investment in learning for children and families, providing hands-on and interactive experiences. Pretty cool, right? After a couple of conversations with the organizers, we came up with the idea of hosting a Farm-toFourth fundraising dinner (the future address of the museum was on 4th Street) right out in the streets of Chillicothe. I received a free ticket to attend and schmooze with the upper crust of society and to talk about my farm. Why? Because I had donated a whole hog to the effort, which was part of the main course! It was an amazing evening by itself, but it also represented a treasure trove of beneficial social capital and exposure to a group of people who didn't know my farm existed.

I hope that the interconnectedness of these stories is not lost on you. Going out and meeting people where they are, speaking and presenting to groups of all kinds, and participating in your local community pays—sometimes in sales, sometimes in connections, sometimes in exposure and marketing, sometimes in social capital and good will—but in my experience it always pays. Talking to a group of five can result in an invite to a group of a hundred, a conversation can turn into an invitation to participate in something special. That is the value to your farm, even if it takes three years for that seed that you plant to bear fruit. So get out there and take advantage of the opportunities available to you to share your story with others, talk about your available products, and connect with your community.

Farm-to-Fourth menu, with my products listed

Let Your Customer Come to You

Even more important than making the effort to go to prospective customers is allowing them to come to you. In my experience, this concept is hands-down the most effective way to secure loyal customers for the long term. It is not without effort or risk; however, I firmly believe that the rewards are 100 percent worth it. "Agritourism" is a topic both broad and deep, so the options and information presented here are less of a complete list for you to choose from and more an effort to get your mind thinking about the ideas, concepts, and possibilities available to you.

A relatively new concept, the term "agritourism" broadly means any agriculturally based activity that brings visitors to a farm or ranch. It speaks to the innate desire that we humans have to connect with something meaningful, that calling that everyone feels for the peaceful and pastoral environment. It implies an intentional welcoming by farmers and ranchers to the community around us, at times of our choosing, to share with them the amazingness that we get to experience every day. The connection that is fostered to your farm and your way of life is one of the most powerful forces available to you to engender trust, secure sales, and grow your farm business.

What this looks like in practice is totally up to you! Some farmers take the approach that customers may show up on any day at any time. I would argue against this approach personally, but if you have the manpower (and willpower) to receive unexpected guests, then great. I absolutely prioritize opportunities to bring folks onto the farm, but always within the bounds of my schedule and calendar—mostly for my own mental wellbeing, as well as to ensure that what needs to get done at critical times actually gets done. With that said, here are some versions of agritourism that have been successful for me.

1. On-farm Pickup

While it is not the first thing that comes to people's minds, on-farm pickup of products is absolutely a form of agritourism. Think of it from your customer's viewpoint: a special trip out to the farm where their food was produced, outside of the normal grocery run that already happened this week. An opportunity to speak directly with their farmer and ask how and what they are doing these days. A chance to visit a physically beautiful property where something important enough to justify the special trip is happening, and a chance to experience just a little bit of that importance.

Hear me, friend: That. Is. *Powerful.* Do not underestimate even the most simplistic of visits.

2. Processing/Harvesting Day

Poultry processing is permissible on-farm in Ohio, under a certain number of birds per year, and so since the farm's inception I have harvested and packaged all my own chickens and turkeys.

After the first couple years, once I felt like I had a handle on things, I went so far as to invite customers onto the farm to participate in processing their own meat. Talk about opening yourself up to a risk, but I wanted to show folks that my process was safe, humane, and clean. In the end, I only had a couple customers take me up on the offer, but I guarantee that they had a stronger connection to their food after they left! The same would be true for a vegetable operation that invites people out to gather their own CSA box each week, or for a u-pick fruit farm.

3. Farm Tours

Two Saturdays per month, in the afternoons, I would offer farm tours. They were scheduled and advertised, RSVPs recommended but not required, and rain or shine. Whether one person or ten showed up, I would spend two to three hours explaining everything from my fencing system to my food safety requirements. We would walk or hayride (depending on animal location) to see every living animal on my farm. We would feed pigs, move chicken tractors, and rotate the cattle/sheep flerd. Then I would take them to one of the high points on my land, overlooking the farmstead, and we would discuss whether or not local and regeneratively raised food could indeed feed our nation. Then we would return to the parking area where I would have my market tent set up, and I would offer to sell any of the products that I had available that day.

This represented about two hours of extra work each tour, both in preparation (signs, setup, hay wagon) and in cleanup. The actual tour wasn't really an extra time commitment—essentially just my normal daily farm chores, but slower and with more talking. I never charged a dime for any of these tours, and found that the sales that I secured far outweighed a per-person fee. The biggest threat was the risk of injury, as folks who were likely more sedentary than I was on a daily basis all of a sudden decided to spend a hot afternoon walking and riding around an environment rife with danger! I only had one fall and one heat-stress incident in my seven years of giving tours, and no lawsuits, thank God. Even though I (as you should) carried a $1 million liability policy, that is not a situation I would ever want to go through, and I would be remiss not to at least mention the additional risk for litigation that you open yourself up to with agritourism. More importantly, and worth every bit of the risk to me: customers left with a better understanding of, and deeper appreciation for, what I was doing out here and what it meant for me to be their farmer.

4. Farm-to-Table Dinner

If you really want to step up your agritourism game, present a polished perspective of your farm, and spend a ton of money, then perhaps a Farm To Table dinner could be a consideration! For my dinners, I wanted to highlight other local businesses, farms, wineries, chefs, and musicians. The tables were set with care, I had friends working as everything from parking attendants to servers, the food was fresh and local, and a farm tour was included as my "staff" reset tables and chairs between the dinner and the outdoor evening showing of a food documentary. I charged $100 per plate, had an amazing time, and lost money every dinner. Was it worth it? I'm assuming so, but most of that investment in time, energy, and money has yet to show up as a return at this point!

The key to financial success for a Farm-to-Table dinner is reaching the audience who can and will pay the (relatively) exorbitant price to attend the event that you envision. I could have cut costs with paper plates and plastic forks, but I was striving for excellence and class. It's just that my local populace didn't see it that way. Bottom line, just like any business, you have to give the people what they want at a price that they can afford. I would venture to say that my failure to do so is why I stopped hosting dinners on my place.

Regardless of how you choose to invite your customer base onto your farm, it is clear to me that you need to do so. The connection that is enabled and secured through agritourism is far too strong to ignore, and it is the most effective way that I have ever found to practice transparency and strengthen the relational connection between your customers and your farm.

Hard work? Yes.

Risky? Yep.

Worth it? Absolutely.

People want—no, need—to connect with their food and with their farmer. When you invite them to come and see for themselves, you feed their inner desire for that connection, engendering trust, securing sales, and growing your farm business. That really is the bottom line of this chapter, and it represents the power available to you as an agricultural entrepreneur. You believe in your method of raising animals or vegetables, in the care that you take for your livestock, plants, and land, and in the environment that surrounds your farm. Ultimately you need your customer to believe in you, too. Give them the freedom to evaluate, challenge, comment, question and engage with you, as you encourage them to take ownership of one of the most important things they control on a daily basis: what goes in their body. Go meet your community where they gather, as well as letting them come see for themselves how their food is raised by visiting the farm, no badge or space suit required. You're proud of what you are doing on the farm, so look forward to the opportunity to show it off to both potential and existing customers. The power in transparency allows you to replace "Animal Welfare Approved" with "[insert customer name here] Approved"!

Write Your Own Story

Create a bullet list of the most compelling, impactful, and powerful aspects of your personal journey toward farming. Then give this list to a friend or family member and ask them what you've missed (we often struggle to see the strength within ourselves).

Write the first draft of your elevator speech, then edit and hone it with the guidance listed in this chapter.

Make a list of local organizations that might be willing to allow you to share your story at their next meeting.

CHAPTER 11

Farm Finances and Budgeting

Before I started my farm, I was making good money flying military airplanes. More importantly for this discussion, my paycheck magically showed up in my bank account on the 1st and 15th of every month, like clockwork. I didn't have to think about it, so mostly I didn't. I was blessed to be able to buy anything I needed and most of what I wanted without truly considering money. What a gift that I took for granted at the time! After leaving the Air Force, I began to make good money raising amazing, clean, healthy food for folks, but something had drastically changed. No longer was I getting paid on a regular basis, but instead was subject to the realities of a seasonal income. The farming money that I made over the summer was significant, but I had to repeatedly remind myself that, unlike my military income, it had to last me all year. This change seems common sense now, so I'm almost embarrassed to say that at the time it completely caught me off guard and required a significant shift in how I managed my finances. Perhaps more importantly, it required that I become intimately familiar with my financial needs, and actually get a handle on how much money I needed to live over the course of a year.

Expense Capture and Budgeting

A farmer's year is broken into seasons, and, as Ecclesiastes 3 says, "for everything there is a season...a time to plant and a time to harvest." This rhythm is comforting and keeps a farmer's work nicely varied, but it is also a challenge in this day and age to make sure that financial ends meet. The first thing you have to know about tackling a seasonal income is how much you spend. I cannot overemphasize this: unless you can say with a high level of certainty, "My living expenses are $XX,XXX per year," then you have some work to do in order to capture that information. If you don't know

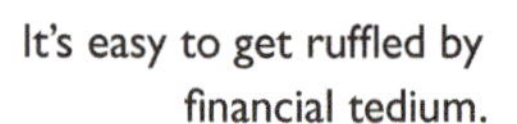
It's easy to get ruffled by financial tedium.

what you need to budget for, then what's the point? Don't feel bad if you don't have that number on the tip of your tongue; I didn't either when I started. I needed to capture my expenses, and did so with a very simple spreadsheet.

Capturing your expenses is easiest done on a month-by-month basis, as you pay your bills, etc. It is possible to accomplish by looking back over an entire year, but that gets sloppier and less accurate. I mean, who really remembers what you bought at Walmart that time over nine months ago? If you have the luxury of taking longer to accomplish this critical task, then by all means take it. If you don't have the time, then just do your best—a 90 percent solution here is better than nothing. Your expenses will vary from year to year, definitely in terms of the amounts but also sometimes with the entire expense. This is even more true as you consider the complete change in location, occupation, or lifestyle that starting your farm might represent. That variability is where budgeting comes in, which is projecting future expenses based on what you now know from your historical expense capture and adjusting those numbers to account for your plans for the upcoming year.

Don't forget that a budget not only includes projections for expenses, but for income as well. You may not be able to anticipate new market opportunities that might surface in the future, but you can reasonably and conservatively guess at what you will have to sell and the value of that product. Use the information in Chapter 6 to guide your income projections by considering your target markets and pricing structure. Ideally, when all is said and done, your projected income will exceed your projected expenses for both home and farm. Don't be surprised if it doesn't, especially early on...it took me three years for my farm to turn a profit, and the average for start-up businesses is five years. The key is that now you know and can anticipate the need for additional income early on, or to reduce projected expenses to a level that projected income will cover. This knowledge represents the true value of capturing expenses and budgeting: allowing you to plan for the challenge of seasonal cash flow.

The last (and most important) step in this process is to stick with your budget. When you sell six grass-finished steers in late July and gross over $17,000, don't forget that you need to buy mineral all year long and that next spring's project will be installing 3,000 feet of new fence line. When you sell your litter of ten fat hogs in October and are staring at $8,100 in cash, remember that you have to purchase weaned feeders in March and be able to feed them for

Proper financial planning ensures that all your hard work has a happy ending.

the following eight months. When you sell your last container of tomatoes, cucumbers, or squash and have secured a tidy profit of $2,600, remind yourself that you need to buy starting pots, soil, and seeds, plus your payment for the hoop house is due every month. If you aren't used to sticking with a budget (which I was not), this concept can be very difficult, especially when opportunity or disaster comes calling. According to Small Business Trends, 82 percent of businesses that fail do so because of cash flow problems. It pays to be regimented in this regard, both literally and figuratively.

Among all of these concepts, the one that I found to be the most challenging was projecting the farming expenses. I now had a strong handle on my family's average living expenses and financial needs, and could accurately predict expense values for grocery, gas, insurance, utilities, and clothing. With a vision and plan in place to market a certain number of animals over the next year, my farm income projections seemed solid and reasonable. But how was I supposed to figure out how much it would cost me to raise those animals? Some frantic internet searching only served to muddy the waters further, as all of the "experts" seemed to disagree and refute each other on the simplest of data points.

Follow me down this rabbit hole of a rant for a minute…

You can't trust the internet farm experts for advice, especially if you are pursuing something outside of industrial agriculture. Very little of the information you find there will have any applicability to you, and will most likely serve to screw you over instead. Here's a perfect example: when I first bought weaned steers in order to finish (aka, fatten) and sell, everything I found online said that they would be finished at 18 months old. Source after source after source said 18 months. It was one of the rare instances where everyone seemed to agree on something! So I bought five weaned calves, put them on pasture, and proceeded to take deposits on my future beef, promising nine customers that their freezers would be full of grass-finished amazingness in less than a year. Raising the cattle was (mostly) a joy, as was the anticipation of putting food on my new customers' plates, but it soon became apparent that something wasn't quite right. As the 18-month mark approached I started to get concerned—the steers did not appear to be growing fast enough. I waited and waited until it was clear that I could delay no longer; not only were the steers not ready—they weren't even close! I was forced to go back to all of those future customers and break the bad news that the beef they were counting on that year wasn't going to happen. Of course I offered to keep their deposits and get them the product whenever it was ready, but only one customer agreed to that and the rest were rightfully upset and went elsewhere. Embarrassed and confused, it wasn't until I shared this story at an regenerative agricultural conference that an experienced rancher explained to me that grain-finished cattle are usually ready in 18 months, while grass-finished animals aren't ready until around 30 months—an entire year extra! I had alienated my first set of beef customers, completely invalidated my financial projections, and endangered my new farm's reputation, all because the internet said 18 months. Now I know that it really isn't the internet's fault, nor is it those who shared that number. They only know what they know. The real moral of this story is to recognize that at least until regenerative agriculture is more widely practiced, most of the information and data that you'll find online does not apply to you.

Except for YouTube videos, those are extremely helpful in almost every instance (how do you think I learned how to castrate hogs?). But I digress.

In the end, I pieced my farming expenses together the best I knew how, but found myself in the uncomfortable position of saying "Well, other people raise chickens/hogs/cattle profitably, so I'm sure I can too"...not exactly the best foundation for a business of any kind. Thankfully I was (mostly) right, and was eventually able to capture my own production data in order to validate and/or challenge my financial assumptions. It took a few years, but once I really honed in on my true expenses and had a much better idea of the market's ability to pay my prices, then my budgeting efforts became much more accurate and valid. This leads me to highlight the need to track data on your farm, something that is necessary but also can be a slippery little slope.

The Value (and Trap) of Data Keeping

As you can see, there is no other type of data that is more valid than that which is gathered on your farm. Someone else's numbers, rules of thumb, and best practices can all provide valuable guidance if you don't have your own information to go on, but your goal should always be to make decisions based on your reality, your situation, and your data. Only then will you truly be in a position to understand where your successes and challenges are, what areas are ripe for improvement, and which enterprises are actually profitable. That is the key word for this conversation: profit. You need to know what your actual expenses and income numbers are in order to determine profitability of an enterprise. Non-financial numbers play into profitability as well; time is one example that comes to mind. Eventually even that data is translated into a dollar amount in order to be inserted into a profit/loss calculation.

If you raise hogs and sell pork, there are some critical questions that need answered:

- How much feed did each hog eat, and how much did that cost per animal?
- How much did each pig cost to source, either as a "feeder" animal or from your sow?
- How much time was required on average for daily care?
- What type of equipment was needed, and what is the amortized cost across each animal?
- What size were the hogs at slaughter? Live weight? Hanging weight?
- How many pounds of meat were returned from each hog?
- What is the value of that product at your current/projected prices?
- How much does it cost you to store and transport meat prior to sale?

All of these questions feed into a profitability conversation but, as you can see, are also specific and unique to your situation. Therein lies the challenge: you need to start with your best guess as you consider starting an enterprise, but validate and update your information after you get rolling with your own data.

How you choose to do this is completely up to you and your personality. I am a spreadsheet guy; it turns me on to drop numbers into cells, use formulas to begin to analyze data, and color code information and conclusions. I know, I know, I'm a weird one! If you are a pencil-and-napkin kind of person, then do that—just make sure that you actually do it. Collecting data on your operation is so important to ensure that your decision-making and profitability conclusions are based on a solid, realistic foundation. Having real numbers to work with will also assist you in the opportunities for accessing capital that we will discuss shortly.

In the hog example above, one of the things I wanted to know was how many pounds of take-home product was I getting back from the processor. So I measured live weight before taking the animals into the processor. I recorded the hanging weight (after slaughter and evisceration). The most painful part of this process was recording every single packaged cut weight before it went into my freezers for storage/sale. I did this for several batches of hogs, after which I stopped because I had what I wanted: an average yield of marketable cuts correlated by weight to any animal running around my farm. I could now answer with confidence when customers asked me "how much meat should I expect" before purchasing one of my animals. I could also project gross and net income, play around with adjusting my pricing, consider the cost/benefit of purchasing feeder pigs versus maintaining sows and a boar, and a host of other critical decision points, now that I had an average yield per hog. Can you see the value in this type of data collection?

The flip side to this coin is that you can absolutely drive yourself bonkers and waste incredible amounts of time recording data that isn't useful. Never should the collection, organization, and analysis of your data get in the way of actually running your operation. Just like anything, it has its proper time and place. Also keep in mind that you don't necessarily need to capture every data point in perpetuity. As long as you can get a solid representation of the information you are seeking, that is usually good enough. Once I had a decent average pork yield I stopped painfully accounting for each individual package and moved on to other things.

Remember that your data is only truly valid until something changes, and change is constant on the farm. If I have my yield data, but then decide to offer boneless loin instead of the bone-in chops, my take-home yield would drop (less the weight of the bones) and my income would increase (more valuable product). Or would it? Guess you'd have to capture some data to answer that question for yourself! What happens when you have the processor package those and you sell them as well—or better yet, value-add them into delicious and nutritious bone broth on your own? You can see how data collection can easily become an unmanageable beast, distracting at best and paralyzing at worst.

The solution to keeping data collection in its rightful place is to make sure that you are answering a specific question that is valuable to you. Then gather your baseline data, make your suggested change, and measure the results. Then move on with confidence in the decision made (or rejected). There are some exceptions to this method—data that I think should be routinely collected simply as a matter of course and good business. Some examples include:

- Expenses: not just the big items that you need for tax purposes, but the little ones that can bleed a business dry before anyone notices.
- Time: regularly accomplishing audits on your human productivity and efficiency is so important to recognizing the need for change and preventing a "that's how I've always done it" mentality from developing.
- Yield: whether in the form of pounds of meat per animal, bales of hay, or pounds of produce per acre, you need to know how much of your end product has been created.
- Sale data: what products are the most popular (and when), which market outlet is most productive, was offering products on sale or bundled together successful?
- Soil health indicators: I deeply regret not being better about taking soil samples regularly from the very beginning, missing out on the opportunity to say "my rotational grazing of ruminants across these pastures has increased organic matter by XX% over the past X years."

Again, all of this data is important for many reasons, perhaps most importantly because it informs conversations regarding profitability. You need to understand and capture real numbers for expenses and income, in order to recognize with clear, open eyes your current financial success, as well as to accurately budget for your farm's future. If you plan to farm for the long term, then the most critical question is: "Are you profitable?"

Enterprise Accounting

Too many times, farmers will oversimplify their answer to that question. They look out over their acreage, think back on the year, and do some quick math in their head. "Well, I made $27,700 and spent $14,000, so yep, the farm made a $13,700 profit." And technically they would be correct, but their self-evaluation doesn't go deep enough. If you only produce one product on your farm, then maybe the answer is close, but with any multi-product farm you need to break apart both the income and expense and apply them as separately as possible to each individual enterprise—hence the term "enterprise accounting." This concept works for any type of farm, but here is an over-simplified example for a livestock operation.

General Accounting Method	
	Whole Farm
Gross Income	
Retail Cuts	$27,700
Expenses	
Feed	$6,910
Processing	$5,820
Seed Stock	$1,270
Net Income	$13,700

As you can see, the farmer is exactly right and they made $13,700 in profit ... so what's the problem? Enterprise accounting seeks to challenge conclusions that are based in generality and to reveal details that are covered over by the broad-brush perspective. It is only when we dig a little deeper and break this example into enterprises that we see what is really going on in this situation.

Now we can see a more complete picture, and we've got some changes to make on this farm! The beef is doing extremely well, due in part to the previous investment in cows that make it so that it isn't necessary to purchase feeder calves, instead raising homegrown

Enterprise Accounting Method				
	Whole Farm	**Beef**	**Pork**	**Chicken**
Gross Income				
Retail Cuts	$27,700	$17,000	$8,100	$2,600
Expenses				
Feed	$6,910	$100	$4,000	$2,810
Processing	$5,820	$2,895	$2,680	$245
Seed Stock	$1,270	$0	$1,000	$270
Net Income	$13,700	$14,005	$420	-$725

cattle. The pork is still profitable, but only just...maybe there are efficiencies to be gained, cheaper feed sources, product pricing adjustments, or bundling of cuts that might either increase gross income or reduce expenses in this enterprise? Most importantly, though, this level of analysis shows that the chicken enterprise is a drain on the cash flow, a fact that was hidden by the profitability in the other enterprises. In this case, removing the chicken enterprise completely would:

- Free up cash flow
- Free up time to spend on more profitable enterprises and general marketing
- Increase overall farm profitability by doing less instead of more

Enterprise accounting provides you with a better answer to the profitability question I asked earlier, where you can answer, "Well, my beef cows made over $14,000, my pork just barely broke even, and I don't raise chickens anymore"!

The value of enterprise accounting is pretty clear, but it is obviously more difficult to accomplish this level of detail. Absolutely worth it, but still more work. On the income side, it is fairly easy to track how much money came in from sales of each specific enterprise, unless you bundle items together. Even then, you just take a percentage of the bundle price and apply it to each individual enterprise.

For me, the real difficulty came in attempting to correctly allocate expenses. For example: all my animals drank water, but they were on the same line that supplied my house, watered my garden, and filled my pool. How on earth was I going to take my monthly water bill and correctly allocate a certain percentage to the farm, much less the individual animal enterprises? For my fencing I used the same step-in posts, twine, and reels for cows, sheep, and hogs. The chickens had their own fencing (netting) but shared the electricity cost of the fencer and connectors with the other animals. The cows and sheep shared a water tank in the pasture, but hogs and chickens each had their own. The cost of my livestock guardian dogs was all allocated to the sheep since they were the sole reason I had them, but what if I decided to sell LGD puppies to other farms and create another income stream outside of the sheep enterprise? These realities of farming reveal the challenge in allocating expenses to individual enterprises.

The answer to this challenge is: figure it out. Make an educated guess. Use this as an opportunity to wield your new-found love of farm data collection, but don't let it keep you from gaining the clear and accurate picture of your collective farm that enterprise accounting provides you!

If you aren't used to considering money this way, the challenges of seasonal income and yearly cash flow can take you off guard. In order to successfully navigate these challenges, the first step is to know what you are spending and make any changes necessary to come up with a realistic annual income requirement. Only then can you properly budget for that income and make sure that your cash-flow projection remains positive year round. When you build that budget, as well as when you capture income and expense information or consider adding a new enterprise, be intentional about taking an enterprise accounting approach to those numbers. This will help prevent unprofitable enterprises from being hidden behind the profit of others and will help you make informed and profitable management decisions. It will also arm you with the information needed to access off-farm funding streams that you might want or need to take advantage of.

Operating Loans

Properly building and using a budget will help ensure your cash flow remains solvent over the course of a calendar year, despite the general seasonality of farm income. Another option that is designed to assist farmers with this challenge is an operating loan. I have to be honest: I have never used an operating loan, which is sometimes referred to as a "line of credit." Not once. I had the opportunity to save money while I was in the military, giving me a nest egg to purchase my initial equipment, animals, and supplies with that savings. After it was gone, I sometimes chose to re-invest farm income back into the farm for animals or machinery and always chose to work within my means to grow the business slowly. This approach takes a lot of patience, which is definitely not my strong suit, but it absolutely pays to stay out of debt if you can.

That being said, operating loans have to be part of a conversation on farm financing, if for no other reason than having all the information in front of you to make the best decision for your situation and needs. In most conventional agricultural operations, farm operating loans are a financial lifeline and a routine way of life. Generally speaking, a farm operating loan represents money that can be borrowed to fund the yearly operating costs of farm operations, including buying seed and fertilizer, paying for labor, and repairing infrastructure. By borrowing money at the beginning of a season for all the expendables, supplies, and some infrastructure improvements, then paying it back (with interest, of course) out of that year's profits, operating loans can help a farm operation manage its working capital and maintain adequate cash on hand.

Most of the time, these types of loans are limited to yearly expenses, meaning that you can't buy a new tractor or other large asset with the money. They are intended to front the cost of operating for the season. The amount of operating capital required can vary significantly on an annual basis, with fluctuating input costs depending on the type of enterprise, but most banks have annual estimated expense budgets available to help farmers determine their estimated input cost based on their plans for the upcoming year. Often these types of loans are ultimately subsidized through the United States Department of Agriculture (USDA) in a form of government support that big agriculture has grown to rely on.

Interest rates vary, of course, but typically hover around 3-5 percent depending on your credit score, pre-existing debt and level of collateral. Speaking of which, don't ever forget that the bank's job is to make sure they are covered in the event of non-repayment! In this case, operating loan collateral typically includes the investment in the current crops grown and the accounts receivable associated with harvest of said crops. In addition, most banks will require additional collateral such as livestock, farm machinery, and equipment, as well as any other farm assets owned.

It is worth saying again that this isn't "free money" or something to be taken lightly. When you involve a bank in your farm's finances, you open yourself up to additional oversight and give away ownership of your operation, to a level that I personally found unacceptable. If you can find a different way of financing your operation or working within your means, I would strongly recommend that. However, there are also plenty of sustainable farmers out there who successfully use this type of loan to solve the persistent cash-flow problem that farming presents. That's the beauty of farming—the decision is all yours!

The Government Is Here To Help

In addition to supporting large-scale chemical agriculture through operating loan subsidies, crop insurance support, and a host of other crutches for that broken system, the USDA theoretically also works for us little guys in a multitude of forms. Most of that support is channeled through either the Farm Service Agency (FSA) or the National Resource Conservation Service (NRCS). In my case, this was a Jekyll and Hyde scenario that I hope and pray is not the norm, yet fear likely is. Keep in mind that this is one man's story and that yours may be totally different. In so many ways, I truly hope that is the case.

Let's start with a cautionary tale, featuring a strapping young lad fresh off his military career who earnestly wants to be successful in his new farming adventure. OK, so the strapping part might be a stretch, but just work with me…it's my story to tell. Anyway, this farmer reads online and is told by those in the farming community that the FSA has a specialized loan program that seems custom made for him: a farm ownership loan targeting (among others) veteran and new/beginning farmers with extremely loan interest rates that would allow him to purchase an adjoining property and expand his rapidly growing operation. How cool is that? So this farmer applies.

The first speed bump in the process came quickly, when he was told that he didn't qualify for the more advantageous "Down Payment" loan because his existing property was larger than 30 percent of the county average, despite the fact that 111 acres is by no means a huge farm to raise pastured livestock on. Unfortunately, he was being compared to all types of farms that don't need as much acreage to operate on, including CAFO-style feed lots, market gardens and backyard produce growers. Despite this unfair setback, the farmer presses on, applying for a "Participation" loan in which FSA acts as a guarantor with a commercial lender to fund a loan. After multiple trips to the loan officer, whose office was a one-hour drive one-way, and many in-depth conversations trying to get her to understand his farm's pricing advantage and operational limitations (no spraying or synthetic fertilizer), it appeared that all the details were in order. All the forms were signed and submitted, and all that was left to purchase the property was notification of approval.

Imagine this farmer's shock and surprise when, a few weeks later, the FSA loan officer asked him to make another trip to her office, where she shared that the loan would not "cash flow" and was denied. The reason: she had independently readjusted the numbers on his previously signed form, adding almost $26,000 of fertilizer expenses to his hay production numbers, simply because she did not believe that someone could make hay without it. Instead of recognizing that this farmer had access to animals who would fertilize the land for free, growing amazing, high-quality grasses without commercial inputs, she chose to alter the loan paperwork after he had signed it and pulled the plug on months of effort, expense, and hope. The farmer was left angry, disheartened, and frustrated.

This story illustrates the issues I have had through most of my dealings with the FSA. While my local office staff are all kind and well intentioned, they simply do not understand anything outside of their wheelhouse of conventional, row-crop, chemical-oriented agriculture. We speak an entirely different language and, in my case at least, the gap was far too wide to bridge. The other key issue is one of bureaucracy, recalling that the FSA is the operational arm of the USDA. That loan officer was the sole gatekeeper for loans in my region. I had no other options; she operated with little oversight, and upon denial my only recourse was to submit a letter of dispute, which was subsequently ignored. The rigidity of the bureaucracy forces you to deal with whomever is assigned to your local office and region, and you are at the mercy of that person no matter what.

The same is true for the NRCS, but interestingly I have had great success accessing funds through them to help with farm updates and infrastructure improvements. Specifically, I have been successful in utilizing the Environmental Quality Incentives Program (EQIP) on two separate occasions, and NRCS also offers a Conservation Stewardship Program (CSP) that supports agricultural operators. The key to successfully talking with the NRCS is to recognize that all of their programs must address/mitigate/solve an environmental concern. Available practices include prescribed grazing, livestock watering and fencing, herbaceous weed control, integrated pest management, cover crops, forest stand management, and organic transition assistance, just to name a few.

EQIP is set up to be a cost-share program, with set payments for specific practices adopted or installed over a specified timeline, which is usually three years. So the negative side of that is that you have to foot the bill for 100 percent of any given practice up front, to the government's exact specifications, of course, and then have an agent come out to inspect that everything is up to snuff before payment is made. While not exact, the payments seem to be set up to cover approximately 50 percent of the costs, assuming that you are hiring someone to do the work. This worked out well for me, as my DIY nature took over and often the eventual payments more than covered my materials and supplies.

So, let's just say that you want to ask NRCS for help installing fencing on your property; what does that look like? First and foremost, you have to have an environmental concern. Maybe there is a pond or streambed that is at risk of excess animal pressure or nutrient leeching. Perhaps the ability to rapidly move animals across an area that is at risk for erosion would help you keep your top soil out of the watershed. I'm not suggesting you lie or exaggerate, but I would encourage creativity where appropriate in order to find all of the ways their money can help you save the planet. If you've got a good NRCS agent, they will assist you with the right words to say to turn on the flow of money to your farm! However you get there, just remember that the practices you are asking for must address an environmental concern.

The second thing you need to remember is that when you invite them on to your land to show off your expressed environmental concern, they are automatically looking at your entire operation for other concerns. I wanted fencing to keep my livestock out of riparian areas along my wet-weather creek and a spring-fed, gravity-flow watering system to prevent compacted areas caused by my water wagon set-up (environmental concern addressed...check!). They agreed, but they also asked me to install a heavy-use pad in front of my winter sacrifice area and gutters on my barn to route clean rainwater directly to the creek instead of it passing across manured paddocks before entering the watershed. It was an all-or-nothing offer. Fair enough, and in my case nothing they asked was something I didn't want or at least couldn't deal with. However, in the scenario where they ask you to do something you don't want to do, turn your back on the money and try again another time. Never let a government bureaucrat talk you into something you don't want—better to politely decline and find another way to accomplish your goals.

My attempts to squeeze money out of the government to help me farm couldn't have been any more divergent of an experience. The NRCS folks have been understanding, candid, and earnest in working with me to address environmental concerns on my property. The FSA, on the other hand—specifically the regional loan officer that I had to deal with—has resulted in nothing but disappointment and angst. That truth underscores my previous statement: this is my story, but I hope and pray that it doesn't have to be yours. Just go into those conversations ready to deal with the frustration and rigidity that I experienced, explain your operation and how it is different than their normal, and hope that your bureaucrat will be different than mine.

Crowd Funding

An alternative source for farm funding that I have had incredible success with is "crowd funding." The idea of crowd funding revolves around the reality that it is easier for people to give if the request is for a smaller amount. It is akin to passing the plate at church—everyone gives a little, yet the total amount gathered is large enough to pay the pastor's salary. Instead of going to a bank to ask for $10,000, what if you could ask 1,000 people to give just $10? Well, nowadays you can do exactly that, through the power of the internet. Instead of having to do all the work yourself, these crowd-funding platforms take care of most of the leg work in reaching a global audience with your need, as well as aggregating the donations into a lump sum that ultimately comes to you—for a small fee, of course!

Within the world of crowd funding there are two different subsets: loans and gifts. Kiva is technically a loan, while others like GoFundMe and Kickstarter result in a gift. I'll explain that further in a moment, but this is why I personally pursued Kiva as my crowd funder of choice. I was uncomfortable with the idea that I would be asking for a handout through the other platforms and wanted to be clear that yes, I was asking for financial assistance, but that I would pay it back someday. I don't say that as a swipe at someone who chooses Kickstarter; just sharing the logic for my decisions.

Kiva (*kiva.org*) is a non-profit company that exists to "expand access to capital for entrepreneurs around the world." They facilitate loans up to $15,000 with no interest rate, repaid over 3 years. Kiva has special allowances for farms, allowing the first loan to be capped at $10,000 (instead of $5,000) as well as allowing up to a six-month grace period before starting repayment. You can only have one loan going at a time, but as soon as you successfully pay off one loan, you are able to turn right around and request another, with the second loan capped at the maximum of $15,000. Needless to say, an average of $5,000 of interest-free money per year in perpetuity (assuming success in getting funded) sounds pretty amazing, doesn't it?

Because Kiva is digital in nature, the traditional bank lending approach of "I'll give you money but can take your house if you don't repay me" doesn't work anymore. Instead, they rely on a concept called "social underwriting" that replaces the bank's request for collateral. The theory is that you make your initial request to your inner circle of family, friends, and coworkers—your social network. If a certain number of people demonstrate their belief in you by giving during that initial phase, then you must be legit enough to take a chance on. It also helps to ensure you repay the loan; can you imagine going back to your mother, pastor, or neighbor and telling them that you decided to take the money and run? I have used Kiva twice—once to purchase five additional cows to rapidly grow my herd and once to buy a haybale wrapper in order to preserve forage for winter feed and mitigate weather challenges increasingly present during hay season.

Other platforms take a different approach, helping businesses and individuals get their ideas funded outright. Kickstarter (*kickstarter.com*) is one such platform, targeting creative ideas to make a reality. Their website says that they are a place "where creators share new visions for creative work with the communities that will come together to fund them." While projects typically are artistic in nature, the platform can still apply itself to farming; you would just have to justify the creative nature of your request in order to make it through the pre-screening process. It is also worth noting that Kickstarter requires you to give some sort of reward to your backers, in return for reaching a specific gifting level. Think outside the box...a plaque on your new barn, having a goat named after them, or monthly pictures of their "adopted" livestock guardian dog...farms have lots of opportunities to provide rewards that other businesses do not.

GoFundMe (*gofundme.org*) is very similar to Kickstarter but more open in its prescribed scope and structure. It isn't restricted to creative business-oriented projects, which is why you'll see campaigns raising money for everything from helping pay medical bills to coping with disasters to furthering social justice efforts. Offering rewards are encouraged but not required and, unlike Kickstarter, if you don't make your fundraising goal, you get to keep what you raise. In addition, as of this writing, GoFundMe has waived their "platform fee," so you keep more of what you have been given, although both platforms still charge a "payment processing fee" to execute and consolidate the gifts for you.

The beauty of any of these platforms is that farms are rich in social capital and therefore have a distinct advantage in the crowd-funding arena. Whether or not they shop for your kind of food, the general populous still strongly supports farming and agriculture. The key to accessing that good will is communicating a solid plan that speaks to your intended audience. Spending some time designing the paragraph or two for your project will pay for itself in spades! Imagine yourself getting an email from a farmer asking for money for a piece of equipment, with this description: "I wanna buy a rapper, so i can rap bales for haylage." Between the poor grammar, misspelled words, slang, and extreme brevity—would you lend money to them?

Now picture getting this description instead: "This loan will be used to purchase a 'round bale wrapper,' which allows high-moisture forage to be baled and fermented. This process creates a highly palatable feed, eliminates wastage due to weathering, and would allow me to cut and bale hay on the same day! As climate change continues to challenge farming operations, this represents a way to harvest forage for my growing herd and flock within a much tighter weather window. Storage requirements would be eliminated, I would have the opportunity to rent the equipment to other farmers looking to take advantage of this technology, and, most importantly, my animals would be better fed throughout the winter! This equipment costs $10,300 from my local equipment dealer, so the entirety of this loan would go toward that cost, and I will cover the remaining balance."

There's a big difference, isn't there? I described the piece of equipment I wanted to purchase in layman's terms—why it would benefit my animals and what problem having it would solve for me. I also briefly touched on how I could help other farmers in the surrounding area and shared specifics on where the money would go and how I would spend it. Those are some of the things you'd want to duplicate in your effort to reach funders from around the country and around the globe, allowing them to understand, believe in, and donate money toward your project. Incidentally, my loan from which that description was taken held the title of "Fasted-Funded Loan" on Kiva for much of early 2019!

The power of crowd funding is immense, cheap to access, and by default skewed in favor of agriculture. Whether you choose a loan-oriented platform like Kiva or a gift/reward-oriented one like Kickstarter or GoFundMe, be sure you fully leverage your social capital and general good will by building a solid loan plan and description. This will ensure that whoever ultimately sees your project request—whether that is your own mother, a barista in Birmingham, a medical technician in the United Kingdom, or a hotel clerk from Switzerland (just four of the 117 individual who funded my loan)—your need and plan are clearly and compellingly communicated. I am of the strong opinion that every single farm needs to have a Kiva loan in progress at all times. $15,000 at 0 percent interest is simply too good to pass up.

Private Funders

This final category will appear like a "catch all" for most folks, containing the private foundation and individual funding options that are normally turned to only when the government and banking institutions fail. However, out of all of the funding methods detailed in this chapter, this is where I have had the most success finding financial support for my farm. When I contemplate that reality, it actually makes a lot of sense. It is the private funders who have the discretion to give their money to a business or farm that speaks to them, whose values align with a foundation's, or whose back may get scratched in return someday with a business deal.

I will describe the private funding sources that have provided amazing support for my farm, but they probably won't be the exact ones that work for you. This category of funders tends to be highly localized and highly specialized, which is likely why they also tend to be highly supportive. Instead of the exact organization (unless they do apply to your future farm dream), I want you to take away the overall concepts and then go find these hidden gems in and around your local area and management practices.

Non-profit organizations are typically formed around either

a set of values or a production method and have money to give as part of their non-profit charter to further their expressed work. The top two non-profits that helped me with both grant money for projects and with agricultural conference scholarships were the Food Animal Concerns Trust (FACT) and the Farmer Veteran Coalition (FVC). The FVC's mission statement (see, I told you that mission statements were important!) is to "cultivate a new generation of farmers and food leaders, and develop viable employment and meaningful careers through the collaboration of the farming and military communities. [They] believe that veterans possess the unique skills and character needed to strengthen rural communities and create sustainable food systems." In my case, the FVC gave me a $5,000 grant to purchase a commercial egg washer, and they also maintain the "Homegrown By Heroes" certification.

The FVC has been amazing to me, but I have to be honest (it's OK to have favorites right?). When I consider non-profit organizations who have informed and shaped my operation and, more importantly for this conversation, put both their money and their full effort behind my farm, there is one organization that comes to mind: FACT "promotes the safe and humane production of meat, milk, and eggs, envisioning that all food-producing animals will be raised in a healthy and humane manner so that everyone will have access to safe and humanely-produced food." FACT has done everything from helping me purchase livestock guardian dogs to sponsoring conference attendance, as well as providing mentoring opportunities for me to give back to others. They are small, but they are committed and loyal and I am proud to be on a first-name basis with most within this amazing organization.

Outside of a formal not-for-profit business, there are many private foundations. These organizations typically exist solely to shepherd a large sum of money in order to further an expressed mission or vision. In my case, I have had the pleasure to deal with the Southern Ohio Agricultural & Community Development Foundation (SOACDF). The SOACDF is entrusted with helping to "create and enhance economic opportunities for Southern Ohio's farm families and rural communities" with the money that was given corporately to Appalachian Ohio's tobacco farmers after the settlement with Big Tobacco in the late 1990s. Originally intended to help tobacco farmers get into a different type of agriculture, there was enough money to go around and I was granted $25,000 to pay for half of a large tractor and multiple pieces of hay equipment—something that I would never have been able to accomplish on my own.

Lastly, a group of private funders that cannot be left out: your customers! Who else would be more inclined to assist you in a specific need, and who else has more at stake in your ultimate success than your customer base? Your local supporters want—no need—you to succeed, and they have already built a relationship with you and your farm products. It should go without saying, but this funding territory needs to be relatively safe...this isn't the place to ask for money to get rich investing in oil futures. Be able to justify a return on their investment for your project, showing that you did your homework and have fully thought your request through, but don't be afraid to ask those closest to you and your farm for their help in making it a continued success.

These are my private funding success stories, but more than likely they may not be yours. However, I have no doubt that, tucked in around your location, management practices, and core values, exists a multitude of non-profits, foundations, and individuals who stand ready to support you in your time of need. Whether it is an unexpected emergency, an opportunity to expand or diversify, or a sound investment in your future success...I have found private funding entities to be some of the most responsive, valuable, and committed organizations. Truly, I'm not sure my farm would have survived, and it definitely would not have been as successful as it was, without their support and assistance.

You may already have this figured out, but I needed to drastically change the way I considered money after I started farming. No longer was that paycheck going to show up magically on the 1st and 15th of each month. Instead I was thrust into the reality of a seasonal income that farming requires. To thrive within that new reality, the importance of accurate budgeting informed by actual expense information and real-world data collection cannot be understated. Gathering that data and allocating it correctly to specific enterprises will help you best understand your financial strengths and weaknesses. This will not only help you make informed decisions within your farm operation, but will also help you speak compellingly to any of the external funding sources you might chose to pursue.

Just make sure that one of those sources is a Kiva loan—seriously there is almost no downside!

Write Your Own Story

Sketch a monthly calendar that includes estimated income and expenses. Begin to identify gaps in seasonal income and brainstorm ideas on how to fill those gaps.

Identify specific enterprises, identify those with shared expenses, and consider options for allocating portions of those expenses to each enterprise.

Write the following statement ten times: "I will apply for a Kiva loan immediately!"

CHAPTER 12

The Business Plan, Part 2—Getting Strategic

Believe it or not, all of the information presented throughout this book needed to be explored and considered before we could tackle this portion of the business planning process. We have finally arrived at the point I promised back in Chapter 3: "Ultimately, you will need to coalesce all of these ideals into a clearly communicated and laid out strategy that will both serve as a roadmap for turning your dreams into reality as well as a way to communicate and convince others to help you along the way." Seems like a long time ago, doesn't it? Nonetheless, since then we have spent a lot of time covering the many avenues available to you as you build each part of your farm business, including branding, marketing, budgeting, and business structures, to name a few. Now, you will begin to strategically apply those options to your vision, mission, and goals from Chapter 3.

Strategy is simply defined as a plan of action designed to achieve a major aim. Personally, I really like that definition because it indicates movement, progress, and action. The strategies that you will develop in this chapter represent a group of realistic actions that, when executed well, will accomplish your goals. This is where the rubber meets the road and all of your hard work forms itself into a legitimate plan. Think of it this way: in Chapters 1 and 2 you tested the farm waters with clear and open eyes; in Chapter 3 you dreamed about and envisioned the end result; in Chapters 4-11 you explored different areas that may help you along the way. Now it is time to create the path that will get you there.

To do that, we will break down the overall task into the same four categories that we used previously: Marketing, Operations, Human Resources, and Financial. This time, however, instead of answering the question of where you are based on your actual experience, you will be making educated and informed guesses about the future, based on your research. This will be a lot of work, but you will reap the benefits of exploring each category in depth, allowing you to make the best-informed decision possible and selecting the most advantageous strategy.

There is one more concept I need to introduce before we can truly be strategic in our planning process: SWOT. No, this doesn't have anything to do with bulletproof vests and semi-automatic weapons! SWOT stands for:

Strengths—Characteristics of the plan that give it an advantage over others

Weaknesses—Characteristics of the plan that place it at a disadvantage relative to others

Opportunities—Elements in the environment that the plan could exploit to its advantage

Threats—Elements in the environment that could cause trouble for the plan

Adopting a SWOT viewpoint as you develop your strategic plans will ensure you have considered what you are great at, where you are vulnerable, what successes likely await you, and what you are lacking to make your dream a reality. As you can see from the descriptions above, Strengths and Weakness are usually internal within your farm while Opportunities and Threats will normally be external to your farm. The purpose of thinking about SWOT isn't just to identify those aspects but to take action on that information—to capitalize on strengths and opportunities while addressing weaknesses and threats, all to the ultimate benefit of your strategic plan and future success!

Marketing

We will start this process with the Marketing category because I believe it is both the most critical and the most ignored aspect of most farming enterprises. Despite the truth that you can't sell anything without having a customer, farmers consistently seem to forget this and adopt a product-oriented approach. This "build it and they will come" mentality, in which we decide what we want to raise or grow assuming that someone will be there to eventually buy it, puts the cart before the horse. Instead, we should start with a customer-oriented approach that considers the needs of our customer base and then uses our values, SWOT analysis, and overarching farm goals to find a way to meet those customers' needs.

The first step in that process is to identify your target markets. Who is available in your area to sell products to? For direct sales, this could be those who live within a specific distance from your farm, or the average number of farmers-market customers who regularly attend. For wholesale, it could represent the number of restaurants in your local town or buyers at the nearest stockyard or produce auction. Once you have identified your marketing options, you need to identify what they value: price, convenience, farm connection, ecological production, desired volume, or something else. These values are the ones you will have to figure out how to meet.

After determining your potential customer base, ask yourself: what are they seeking? Is there a niche that is not being offered elsewhere? What is it about my farm that will speak to these customers? Keeping the mindset that you will ultimately be producing something to meet their needs will allow you to keep that horse out in front of the cart, helping to ensure that once you do produce that amazing product that there will actually be someone there ready to purchase it! It is also likely that you aren't the only game in town asking these types of questions, so you need to identify your competition as well as your customers. Who are they, what do they produce, and where do they sell are a few simple questions to get you started. This will help you in two ways: first in identifying those niche areas that are not being served among your prospective customer base and secondly in being realistic regarding the percentage of "market share" you can expect to secure.

At this point, you should have at least one idea of a potential customer and a product that they need; likely you have several. The last few considerations are intended to make sure that you can actually meet said need by logistically getting your product to the customer. If your customers are those driving by on your busy road frontage, then the logistics might be pretty simple! However, if your customer is the local grocery store or restaurant who needs a small amount of product on a regular schedule, then that becomes an entirely different consideration. If you plan to sell meat "custom cut," meaning that you drop off a live animal at the processor and your customer picks up their custom-cut, packaged meat directly from them, then you wouldn't have to worry about storage issues. However, if you began to sell individual cuts to retail customers, you would likely need to consider on-farm or rented freezer storage, along with the associated food-safety red tape that almost certainly will come with it. Think through the logistics (and legalities) of your proposed customer-product combination, in order to capture the full set of costs and requirements for you to meet that need.

Operations

It is likely that you already have a strong idea of what your farming operation will look like; this is usually where our minds head whenever someone asks us to describe our future farm dreams. However, you still need to consider this aspect of your business strategically, especially after you just got done putting a plan together on who will buy your products. You probably won't make wholesale changes to your production plan or methods, but you should be mentally prepared to consider alternatives and make adjustments.

Your operational strategy covers more than just methods of production. Especially in food production, there are a myriad of regulations to consider. If you choose to pursue a third-party certification, then that organization will add their own layers of operational requirements. And then there is your production capacity to think through, both in terms of your available land base, time availability, and human-resource options. None of us are superhuman, so we need to make sure that our operational strategy fits our lifestyle requirements and capabilities. This is the nitty gritty part where you finalize your first choice for what, how, how much, where, and when you will produce your farm products.

As you put pen to paper for this section, be as specific as possible and break the strategy down by enterprise (cattle, sheep, carrots, rutabagas, hay, etc.). This will allow you to drill down deep and capture the subtle details that need to be considered. Discuss your fertility and breeding program, fencing, pest and water management, mechanization, and waste control. Describe what went into your decision to process poultry on-farm instead of at an off-site location, or which slaughter facility you decided on and why. If you don't have the specifics ironed out quite yet, then talk about the factors and ideals that will eventually guide and influence your decision. Detail influential conversations you've had with consultants, family members, and fellow farmers. Get deep and capture all of those details that inform your proposed production practices.

Production Schedule

One of the most important things to think through is your production schedule. Farm life is, by default, affected by seasons as much as by markets, and it behooves you to spend some time in front of a calendar. Remember my eighteen-month turned thirty-month grass-fed beef production train wreck? This is your chance to avoid that mistake! My general approach is to start any production schedule from the end goal and work backwards, meaning that I determine when my customer needs a product, then step back to processing, then back to finishing, finally back to the right time to purchase that feeder stock or have those babies on farm. That approach will tell me when I need to source animals in order to fully finish them, process them, and have a product to sell. The same concept works for having seeds and starting pots ready to go so that you can plant, start, harden, and replant your produce in order to have vegetables ready for picking when your markets are open.

While customer demand is one constraint on your production schedule, you would be foolish not to consider seasonality. I always wanted to have my babies arrive on-farm (southern Ohio) in the warming months of April and May. I had fresh green grass available, calving and lambing could comfortably occur outside, and my management tasks were greatly reduced. I always rolled my eyes at my conventional neighbors as they described the sleepless nights, frozen early mornings, and birthing difficulties they self-created by breeding for babies in February! Make sure to consider the weather and seasons as you build your production schedule.

Recognize also that customer demand and seasonality seldom match up exactly. If you plan to sell weaned calves at the sale barn, you may need to calve in January. If you want to target the mild months of April and May for calving, then you won't have a grass-finished product until thirty months later, processing animals in July and August, which is already halfway through "farmers market season." Whichever influence rises to the primary consideration for you, at least you'll be in a position of recognizing the issue early and (hopefully) addressing it well. Whether you are growing carrots or cattle, you'll need to consider the seasons, your markets, and your physical availability in order to ensure that you don't get caught in the trap of growing amazing food only to find out you can't process and/or sell it!

Tools & Equipment

What tools and equipment do you need to accomplish your operational strategy? What do you already have access to, either via ownership, borrowing, external agencies, or contract operators? Think about land, but also buildings, machinery, equipment, and supplies. If you are planning a rotationally grazed cattle herd, then you'll need mineral blocks, water tanks and hoses, fencing reels, and posts. You should probably have pest control, basic medical, and castration supplies on hand. In most of the United States, you'll need some sort of heater to keep water melted and shelter for your animals over the winter months. All of that just to produce grass-finished beef! And the list is totally different if you are planning hogs, lettuce, grain, hay, or mushrooms. As you can see, this task is a robust one for a beginning farmer, but the more effort you make now nailing down these details and the tools, equipment, and supplies you'll need to actually make it happen, the better. This is also where glaring "resource gaps" will reveal themselves, so be sure to allow your available resources to inform your operational plan.

Regulation

Unfortunately, we can't have a food production conversation without also considering bureaucratic regulation. I would never scoff at legitimate food safety concerns, but far too often government regulations step well outside their bounds. That being said, you'll have to navigate that tangled web soon enough, so you might as well get yourself well acquainted with it! Talk to your local health department and state department of agriculture to get you started, it is likely that most of what you want to produce is regulated by one of those two, and they can tell you who to talk to if that isn't the case.

If there is overlap in costs, equipment usage, or infrastructure, do your best to detail and account for them separately, allocating a percentage of the overall cost to each individual enterprise. This will ultimately help you in the final lesson, where you will evaluate your strategies, select the best one, and develop contingency plans.

Human Resources

If you are like me, when you hear the term "Human Resources" you probably picture someone in a suit whose job it is to pour through job applications and conduct interviews for potential workers. Does any of that apply to a small family farm? The answer is yes, even if you are starting your operation by yourself or within a relatively small family unit. There are some things that must be considered for any sized business to be successful: labor (both the tasks themselves and the people who will be accomplishing them), compensation, decision-making responsibility, and communications.

Labor

As you developed your Operational Strategy during the last section, you almost certainly began to mentally assign that work to yourself and others. Now is the time to formalize those assignments, being sure to account for the individual strengths and weaknesses that you identified as part of your SWOT analysis. Consider routine chores, specialized seasonal tasks, and travel to and from markets. Make sure to build in time and personnel requirements for social media marketing, record keeping and third-party certification—you don't want those to be an afterthought. Be realistic when it comes to allocating your time, and consider the innate seasonality of the farming workload. You don't want to pile up too many critical tasks on one person during one season, although that will always be a reality of this lifestyle to some degree.

When you do identify periods of over-extension, you'll need to address them now. Perhaps you can reallocate the work to another team member? Otherwise, you might need to consider bringing in additional help for those busy times and tasks. Do you have family close by? What about other farmers who might be willing to trade assistance in times of need? Without those options, you may need to consider hiring either seasonal or full-time labor. Personally, I avoided hiring employees, even though I definitely could have used the help, because of all the additional red tape that came with it. In many cases you have to offer insurance, contribute to workers compensation, and deal with a host of extra rules. Often, the best way to handle the extra workload is to hire custom operators to accomplish specific tasks (make hay, transport animals, deliver product, etc.). This has the benefit of getting the job done without any additional tax implications and can help you avoid costly equipment purchases; you just need to watch your cash flow to ensure you keep the money set aside to pay for the services you hire out.

Compensation

Speaking of paying for services, don't forget that you need to pay yourself! Farmers are the worst at this, falling into the trap that working close to the land or contributing positively to feeding those around us is somehow compensation enough. Those are amazing things, but absolutely do not count them as repayment. If you want to continue to enjoy and provide those things, then you must insist on paying yourself at least a living wage. That being said, you also have the ability to be creative here. One example might be to trade our time for our products, assuming you like to eat whatever it is you produce! Instead of selling your products for cash and then spending that cash at the local supermarket, just keep your money and eat your own products.

A word of caution is warranted here: I know that sounds like a really simple concept, but where I see it messed up most often is by not actually tracking the value of the packages that cross that line between the market cooler and your personal refrigerator: they need to be accounted for as business income as well as a personal expense. If you simply help yourself to your market products without treating it as a transaction, several things get missed:

1. Your enterprise accounting will be incorrect, as the value of products that came from your acreage or animal just disappear without the corresponding sale data that should come with it. Essentially you can no longer accurately calculate your net return per acre/animal, nor can you validate your individual product pricing.
2. Your personal budget will be skewed, as you won't be accounting for the actual cost of food you need to support yourself or your family, if for some reason you would need to purchase them sometime in the future.
3. It is far too easy to bring a bunch of meat to the family's annual 4th of July barbecue out of the kindness of your heart, or have a few Thanksgiving turkeys end up on your own plate, without actually getting paid for the time, effort, and expense that you went through to produce them.

I'm not saying "don't share"! By all means, eat your own products. What's the point of producing amazing food if you can't enjoy it yourself? However, to address these concerns, I recommend that you actually purchase your products from your farm business with your personal money. If you'd prefer to avoid that formality—say for taxable income purposes—then at the very least you need to track these transactions and make sure it is accounted for within your internal calculations. That way you maintain the validity of your data, you avoid the trap of giving away too much, and most importantly you keep the reality that you need to pay yourself at the forefront of your mind.

That is just one creative idea regarding compensation. I'm sure you can come up with several others. Regardless of how it happens, I cannot emphasize it enough: please be intentional about paying yourself for your good and important work!

Decision making

A separate, yet equally as critical, piece of the Human Resources puzzle has to do with decision making. Hopefully you have surrounded yourself with a team of committed individuals throughout this process, in order to best inform and shape your future farming enterprises. That team shouldn't disappear once your farm dream is a reality, but the truth of the matter is that you also can't manage the day-to-day operation of a farm by majority vote. Once your strategic plan is in place, someone has to execute that plan with boots on the ground. That person could be you, but it also doesn't have to be. The bottom line is this: someone (or a very small team of someones) needs to be empowered and charged with the ability to make operational decisions for your farm, and that person needs to be specifically identified in your plan. Different people can be decision makers for different aspects of your operation if that makes sense to you, but those roles and responsibilities should be identified in your Human Resources strategy.

Communication

One of the bedrock requirements for success in any partnership, be it a marriage, friendship, or business, is communication. While effective communication won't (and shouldn't) prevent disagreements, it absolutely will mitigate and soften them, while poor communication will almost certainly ensure the worst-case scenario.

Daily planning briefings, weekly team sessions, or quarterly membership meetings—however you choose to communicate with your team, you'll need to identify how (and how often) you intend to keep your team in sync and on track together.

Financial

This category of strategic planning tends to be an overflow of the previous three categories. Perhaps more accurately, your strategic plan for money likely stems from and influences your strategy for marketing, operations, and human resources.

In Chapter 5, we discussed the different options for structuring your farm business: Sole Proprietor, Limited Liability Corporation, etc. That decision will need to be identified in this strategic plan, not only because of the money that will be required to set up the structure, but also for the tax implications, fee structures, and compensation plans.

Additionally, plans regarding land access influence your rent and/or mortgage payments, projected enterprises will generate start-up and capital investment costs, and operating expenses need to be projected forward. You'll want to revisit Chapter 11 for those thoughts, as well as to detail any loan programs or grant opportunities you are anticipating pursuing to assist you in funding your farm dream. All of these decisions and insights, specifically their financial needs and implications, need to be identified within this section.

There will be lots of unknowns in this section as well, especially if part of the reason you are accomplishing this business plan is to support your application for grants or loans. If that is the case, remind yourself that the strategic plan is your "best guess" for the way forward, and all of this information is less of a concrete decision and more of an initial plan for execution that is subject to review and reshaping.

Putting It All Together

Now is the point where we combine the marketing, operational, human resource, and financial elements into a fully integrated whole-farm strategic plan. This will represent your first choice for individual strategies, as long as they are all compatible. The struggle in pulling them all together will be when there are competing interests and desires, which is where your values and goals from Chapter 3 need to come into play. Think about things from an overall, systemic perspective.

Once you have your primary strategic plan in place, the first thing you need to do is take a break! Up to this point in this chapter, you have been doing some really heavy mental lifting. Seldom will an author tell you to put their book down and walk away, but in this case it will be extremely healthy to step back and take a few days or weeks to give your brain a break. This will help clear your head, get you out of the mental rut that you are undoubtedly guilty of settling into, and allow you to fully and accurately consider alternatives to your primary plan. Just make sure to pick the book back up and continue after you've taken the chance to reset!

After your hiatus, the most important test for your strategic plan is to go back to your values. Does anything that you've written go against them? You'd be surprised how many times things can sneak their way in without us noticing them, so make sure that your values are accurately and wholly reflected in your strategy. A strategy that flies in the face of, or skirts around, your values is worthless. Your values must drive and support the strategy.

The next test is against your goals. All of your expressed goals don't have to be represented, but at least your prioritized goals should all be present and accounted for! This is the step that makes sure that your concrete plans are in line with your dreams, which informed and shaped your goals. Consider this a "quality of life" test, which will serve you well once you are living the farming life in person that is represented on paper.

Lastly, take one more look at the financial section of your strategy, in the spirit of a risk evaluation. Before you begin to spend your savings account, or invest your future off-farm income, or stand in front of a lender with your amazing ideas, you need to double-check that everything still pencils out after considering your best marketing, human resources, and operational plans. Make sure that the volume you can realistically produce, sold at the lower end of your price range and including actual production, processing, and storage costs, covers your cost of production (including paying yourself, of course). Does your seasonally received projected income spread itself out over the entire year of your projected expenses? In other words, does your "cash flow"? If not, why not, how long will it take to answer "yes" to that question, and is that acceptable to you? What happens to your plan in the face of adversity—say, a global pandemic or something of that nature?

These tests are all important, but they certainly don't cover all of the possibilities. Feel free to create your own tests in addition to the ones I mentioned here, whether they be environmental, social, or personal. If you find disparities, make the necessary changes (or accept the existing plan with its limitations), and then re-run your tests. Only then are you ready to complete this process by writing your business plan!

Writing Your Business Plan

I'm sure this has been a challenge, but I promise that it will be totally worth it in the end. Now is the time to consolidate all of your values, vision, analysis, and strategy into one formal document. Don't let off the gas yet—the job isn't done until you are able to honor all of your hard work by effectively capturing and communicating your research and vision for your farm.

The actual formatting of your finalized business plan can be variable, depending on your intended usage. If you need the document to be externally facing (i.e.: lenders, investors, or potential partners) then you'll want to emphasize your strategic plans, market research, and pricing decisions. If the plan is more internally facing (team members, operational plans, etc.), then your values, vision, mission statement, and operational strategy will be at the forefront. Regardless, the formatting ultimately needs to make sense to you and clearly capture your intent and decisions made throughout this process.

With that in mind, a general format should probably look something like this:

- Cover
- Executive Summary
- Table of Contents
- Values Statement
- Historical Context
- Current Situation
- Vision Summary
- Mission Statement
- Prioritized Goals
- Marketing Strategy
- Operations Strategy
- Human Resources Strategy

- Financial Strategy
- Appendices (as required)

I recommend a simple word processing document, allowing you to update and change data as you receive feedback. You can add images, charts, and graphs easily and can export the document as a PDF to save on space and prevent unauthorized editing by those you share the plan with. Each section can have its own page, or you can separate sections with a centered header. Font and text size is flexible, as long as you stay "normal." (This is not the time to give the WingDings font a try!)

A word on the executive summary: from a communication perspective, this is the most important part of your business plan. It is the "gist" of your plan, concentrated and distilled down to a page or two at most. It is the section that everyone will start with, and it might be the only section that someone will read! That isn't meant to discourage you regarding all the effort you have put in to the entire document—only to encourage you to take the executive summary seriously. Because of its brevity, it is also the most difficult section to write. You need to strive for clear communication of the need and/or opportunity, your strategy for accomplishing your goals, and your excitement and preparedness for tackling the project in front of you—a tall order indeed for a max of two pages!

A good way to tackle the executive summary is to take each section of the plan mentioned above and turn each of those into one or two paragraphs. After the normal identifying information (name, business, location, etc.), pull mainly from your history, current situation, market research, product offerings, and the four areas of strategic planning. Do your best to pull out the clearest, hardest-hitting data and match it against the market opportunities that are driving your production decisions. The executive summary needs to be clear, concise and positive yet realistic, and it should compel someone to read deeper into the specifics of your business plan.

With all of this feedback in mind, it is time to write your full business plan, assembled from all of the previous tasks from this book. Enjoy watching your history, values, vision, and goals formally meet your opportunities, production methods, and strategic plans—your farm dreams are so close to becoming a reality!

Implementation

Now that you have your finalized business plan, it's time to clear off some shelf space, right? NO! This plan was built for action, for execution, for progress—not for your coffee table! While there is undoubtedly value in taking a deep breath as this process comes to a close, it would be easy to stall out there if you aren't careful. There is a very real difference between talking about something and doing it, and it is possible that you may feel inadequate in your preparation. Don't lock up into inaction. While you will never know all the details or have every part nailed down, you should have more than enough to move forward confidently with execution of your business plan.

I love a good list. In fact, occasionally I have been known to make lists of lists! OK, it is a daily occurrence for me—some might say a sickness. But in this case, a list of tasks, and who is responsible for accomplishing tasks, will assist you in taking those "small bites out of the elephant" in order to make the overall task more manageable. Take your highest priority needs and first steps, write them out with specific tasks to get them done, and assign them to someone on your team. Be sure to include a deadline; it is amazing what clear expectations for accomplishment can do to ensure success in moving forward towards a goal. When one task gets completed, revel in that progress and then figure out what other task was waiting on the first one to be finished and add it to your list. In this manner, you will quickly, efficiently, and accurately move your business plan from paper onto your property, which has always been the primary goal.

Years ago, as I was being trained for my first career as an Air Force officer, we often discussed Colonel John Boyd's concept of the OODA Loop. John Boyd was a military strategist who used the OODA Loop to describe the process of decision-making by either an individual or a collective. Besides being just plain fun to say three times fast, the OODA Loop breaks down the decision making process into four parts that end with action. The fact that I want to highlight here is that the process does not end with Action, but instead reconnects in a continuous loop that follows Action by a second Observe. In this way, decision-makers Observe the results of their Action, re-Orient their viewpoint to the newly created reality, make a Decision regarding progress along their desired path, and (if needed) Act to ensure their reality is heading in the right direction before Observing again. You get the picture...

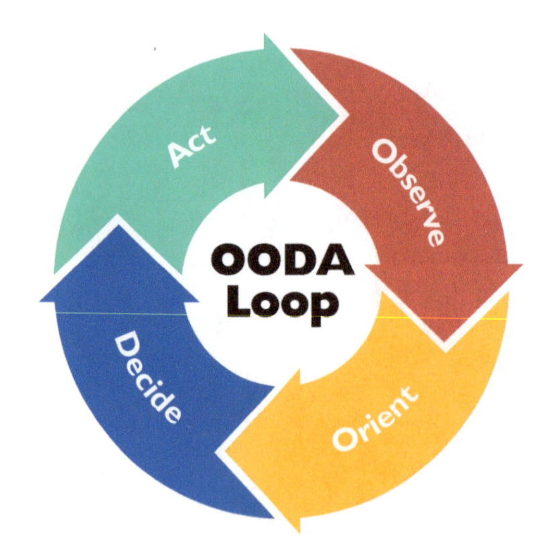

Observe
What is the current situation? What is the reason you want to change? How bad do you want to change?

Orient
Where are you currently at relative to where you want to go? How far is it to your destination?

Decide
What is the exact path you are going to take? How are you going to handle challenges and set backs?

Act
What is the approach and method you will take to implement the decisions? What is your action plan?

In the spirit of the OODA Loop, continue to Observe your progress towards your goals. One major way in which to accomplish this is through accurate record keeping. How will you know if you met your first year goal of "breaking even" financially if you don't track weekly sales data? Record keeping is extremely important, but as we discussed in Chapter 11, it can also represent a rabbit hole of wasted time if taken too far. The balance that you'll need to find allows you to track important data sets that will either measure success or failure in accomplishing your goals or provide input into future decisions.

Ultimately, your business plan not only serves as an initial roadmap towards accomplishing your goals, but also a benchmark to measure your progress against. Goals exist to encourage you to move forward, but also to measure success, failure, progress, and hangups. You need to bring the results of your data collection and record keeping back around to your business plan for reevaluation, continuous improvement, and your next iteration of your farming OODA Loop. The bottom line is that your business plan should be an active part of your next steps. Use it to encourage, use it to inform and equip, use it to educate others on your intentions. Use it to make your farm dream a reality.

Now... go do it!

Index